TECHNOLOGY,
R & D, and the
ECONOMY

TECHNOLOGY, R & D, and the ECONOMY

Bruce L. R. Smith and Claude E. Barfield
Editors

THE BROOKINGS INSTITUTION
and
AMERICAN ENTERPRISE INSTITUTE
Washington, D.C.

Copyright © *1996*
THE BROOKINGS INSTITUTION
1775 Massachusetts Avenue, N.W., Washington, D.C. 20036

and

AMERICAN ENTERPRISE INSTITUTE FOR
PUBLIC POLICY RESEARCH
1150 17th Street, N.W., Washington, D.C. 20036

Library of Congress Cataloging-in-Publication data

Technology, R&D, and the economy / Bruce L.R. Smith, Claude E. Barfield,
eds.
    p.   cm.
    Includes bibliographical references and index.
    ISBN 0-8157-7986-0 : alk. paper — 0-8157-7985-2 (pbk. : alk. paper)
    1. Research, Industrial—Economic aspects—United States—
Congresses. 2. Technological innovations—Economic aspects—United
States—Congresses. 3. Research, Industrial—Economic aspects—
Congresses. 4. Technological innovations—Economic aspects—
Congresses. I. Smith, Bruce L. R. II. Barfield, Claude E.
HC110.R4T433   1995
338′.064′0973—dc20   95-47505
                                              CIP

9 8 7 6 5 4 3 2 1

Typeset in Sabon

Composition by Harlowe Typography Inc.
Cottage City, Maryland

Preface

THIS VOLUME presents the papers and commentary emerging from a conference undertaken jointly by the Brookings Institution and American Enterprise Institute for Public Policy Research and held in October 1994 under the sponsorship of the National Science Foundation. The purpose of the conference was to analyze the contributions of research to the economy and to society. The authors have substantially revised their original papers in the light of the conference discussions.

The aims of the project were to assess the role of R&D in the economy, to identify promising new areas of research and analytical approaches, and to contribute to the public debate in a broad sense as the nation seeks to define a new framework for its R&D policies in the post–cold war era.

Since our two research institutions have been engaged in studies of R&D policy for several decades, the collaboration seemed a happy convergence of intellectual challenge and institutional tradition. The contributors brought analytical rigor to the endeavor. We thank them for their gracious approach to the hard work required of the project. Our sponsors from the National Science Foundation were unfailingly considerate and helpful while we struggled to bring the project to conclusion. We are especially grateful to Kenneth Brown of the National Science Foundation for his encouragement and wise advice throughout the effort.

Many staff members from AEI and Brookings assisted at various stages of the project, and we regret that we cannot acknowledge them all. A few must be singled out for special thanks, however. Isabel Ferguson and Michele Van Gilder skillfully organized the conference and handled the complex logistics. Michael Voll served as research assistant and verifier, devoted long hours to reformatting footnotes and complex tabular and graphical material, and generally assisted in preparing the manuscript.

Marjorie Crow provided typing services. Colleen McGuiness edited the manuscript with great care, Carlotta Ribar proofread it, and Julia Petrakis prepared the index.

The individual authors are, of course, responsible for the views and opinions expressed in the papers and comments, which should not be ascribed to the National Science Foundation or to the trustees, officers, or staff members of the Brookings Institution or American Enterprise Institute for Public Policy Research.

BRUCE L. R. SMITH
CLAUDE E. BARFIELD

Contents

CHAPTER 1

Contributions of Research and Technical Advance to the Economy

Bruce L. R. Smith and Claude E. Barfield

In 1972, when the National Science Foundation (NSF) sponsored a colloquium on research and development (R&D) and economic growth and productivity, the field was in its infancy. Since then, a great deal of additional research has yielded paradoxical results: Knowledge of the linkage between R&D and growth has advanced significantly, but awareness of the complexity of the issues and of the remaining gaps has increased. Further, the need for more communication between scholars and policymakers to improve the quality and relevance of the research is more evident than ever.

During the past three decades, analysts have become convinced that innovation is a continuous, disorderly, and complex process, not a discrete event or series of linear events. The solution to a basic scientific puzzle or the invention of a new product in a laboratory makes no direct economic contribution. Innovation includes not only basic and applied research but also product development, manufacturing, marketing, distribution, servicing, and later product adaptation and upgrading. Classification schemes that describe the innovation process as a straight-line progression inevitably fail to capture its essential messiness and serendipitous nature. The process contains numerous interactions among all stages, leaps ahead, feedback loops, and sudden and unexpected lacunae. For instance, in the development stage, problems may arise that reveal gaps in fundamental scientific knowledge. Chance and seemingly unrelated breakthroughs in basic science may, however, open up new avenues for technological applications. This will likely apply, for example, to superconductivity—the end products may range from computing to energy storage and transportation. Finally, users of the new technology may call for substantial redesign of the initial product. The importance of redesign and modification illustrates a significant, and often over-looked, truth about innovation: At any given time, most R&D involves product improvement and technological refinement, not the creation of new knowledge or revolutionary new products.

Although certain factors are central to technical advance in virtually every sector, important differences among sectors affect the nature and the source of technical change in individual areas. For example, in aircraft and telecommunications, the end products are complex systems composed of many subsystems and components. Technological advances may stem from improvements in individual components or from dramatic system-level redesign; in either case they result from the work of upstream component or materials producers and systems engineers. In some systems technologies, users may be especially important (for example, aircraft and the airline industry). The chemical and pharmaceutical industries, however, manifest a strikingly different model. Innovation in these industries is characterized by the introduction of new products, and much less of the incremental upgrading characteristic of systems technologies is evident. Input suppliers play little role in technical advance, but users may play a large role through the discovery of new applications for a chemical or medical compound.

Because of the complexity of the innovation process, determining precisely the qualitative or quantitative benefits to society from individual research projects has been difficult. Although observers strongly suspect a positive correlation between investment in R&D and economic growth, the exact nature of the relationship has remained elusive. Economists have, however, made various attempts to examine the quantitative and qualitative benefits to society from investment in research. Three approaches have been common. First, economists have constructed models attempting to describe the relationship between R&D inputs and such outcomes as changes in the value added, net profits, output, or rate of growth. Various levels of aggregation have been used, from the individual firm to whole industries, regions, or national economies. Among the studies based upon econometric modeling, two types of analyses are especially noteworthy: growth accounting and individual industry modeling. Robert Solow pioneered the field of growth accounting with research on the rate of technological change in the United States during the first half of the twentieth century.[1] Solow concluded that a residual or unexplained

1. Solow (1957).

portion of U.S. economic growth stemmed from technological advances and that this residual far outweighed changes in capital or labor.

Subsequently, other economists refined and broadened Solow's analysis to constitute a second major approach. Edward Denison's work in growth accounting has had a big impact on the thinking of economists and policymakers.[2] Denison estimated that 20 percent of U.S. economic growth between 1939 and 1957 is accounted for by R&D. Edwin Mansfield attempted to measure the payoff from research through the analysis of technological change in numerous industries. In an initial set of studies, Mansfield found that, for the period 1975–85, about 10 percent of the new products in seven industries—information processing, electrical equipment, chemicals, scientific instruments, drugs, metals, and oil— could not have been developed (or would have faced substantial delays) without the use of recent academic research. He found that the mean time lag in these industries between academic research and first commercial product or process introduction was about seven years. His tentative estimate of the social rate of return to society (that is, the benefits to society beyond those to the industry itself) was 28 percent. He singled out drugs, scientific instruments, and information processing as industries for which academic research had been particularly important. Mansfield's chapter in this book summarizes the highlights of this past research and adds new data and analysis from current research.

A third approach to estimating the payoff from R&D consists of detailed case studies tracing the history of individual products or processes to their research origins. The Department of Defense's (DOD) Project Hindsight (1966) and the National Science Foundation's Technology in Retrospect and Critical Events in Science (TRACES) project (1968) are examples of this type of analysis. Project Hindsight explored the origins of specific defense weapons systems and concluded on the basis of a cost-benefit analysis that the DOD investment in R&D in the period 1946–62 resulted in a high payoff for the department in fulfilling its mission. But the biggest short-run payoff came from applied research and advanced development work supported by the Department of Defense. The weapons projects drew only indirectly on the common pool of basic research that was widely available.

The TRACES project ascertained the evolution of five innovations (magnetic ferrites, oral contraceptives, NCRs, the electron microscope,

2. Denison (1974, 1979, 1985).

and matrix isolation) and described the institution's program and R&D activities that produced the successful outcomes. The TRACES project demonstrated the importance of the basic research knowledge base for each technological advance and showed that a relatively quick (ten years) average payoff for the research occurred. The TRACES project, however, produced a warning on the difficulty of determining the scientific roots of any technological advance, not least because of the serendipitous nature of the entire innovation process. This cautionary note is worth recalling in the context of the present debate over the role of basic research in U.S. competitiveness.

The current state of research is thus marked by both progress and continuing intellectual puzzles and gaps in knowledge. In the general context of greatly increased public interest in the relationship between technology and economic advance, it seemed timely to undertake a major review of the subject. Independently, the Brookings Institution and the American Enterprise Institute for Public Policy Research (AEI) discussed with the National Science Foundation the idea of an anniversary conference to revisit the major themes of R&D policy and the economy discussed at the 1972 NSF colloquium. The NSF suggested that Brookings and AEI might fruitfully combine forces in such a venture. We happily agreed and submitted a joint proposal to the NSF for the conference.

The Aims of the Conference

A first goal of the conference was to assess the current state of knowledge and to note the advances made in the intervening years since the initial colloquium. In particular, we wanted to assess the more recent contributions given the heightened awareness of the complexity of the R&D process, the globalization of technology and of financial markets, and the increased international competition in many high-technology sectors. What were the implications of the new findings for public policy and for corporate competitive strategies? But we were also interested in the broader implications of the contributions of research in areas beyond economics, such as education, health, environment, and quality of life. A further aim of the conference was to identify the priorities for additional research on the most important questions.

To accomplish these aims, a broad group of researchers, policymakers, educators, and corporate leaders was convened to review the research findings and to discuss the full range of issues. The need to examine the nation's R&D strategy from a broad perspective appeared both timely and compelling. For most of the postwar period, the nation's R&D effort has been focused on national security, space, energy, and health. Five years after the fall of the Berlin Wall in 1989, policymakers have not fully assessed the implications of the changed national security environment for the size and the nature of the U.S. scientific and technology establishment. The defense R&D budget has been reduced only modestly even though troop levels have declined substantially (and probably will decline even further to levels below the 1.6 million base force announced by the Bush administration or the bottom-up projections of the Clinton administration). Basic changes in the defense R&D system thus appear likely, especially as smaller troop levels put more pressure on the military to maintain a high state of readiness.

New goals have surfaced on the public policy agenda, including the desire to deregulate the economy, combat crime, reduce health care costs, and lower the budget deficit. These objectives may call for redirections in the country's R&D investment as some technology-intensive missions decline in importance. The Clinton administration came to office in January 1993 with pledges to reorient the balance of the national effort between civilian and military research, to augment the resources devoted to various National Institute of Standards and Technology (NIST) programs, to promote "dual use" technology through the Defense Department's Technology Reinvestment Program (TPR), to complete the information superhighway, and in general to link technical activities and research goals more closely to broad social goals. Meanwhile, in Congress, the then-chairman of the Senate Committee on Commerce, Space and Science, Barbara Mikulski, D-Md., was pushing hard for a 60 percent numerical target for strategic research in the total R&D effort of the National Science Foundation. The nation's continuing fiscal crisis made more likely further pressures for budget cuts in the area of discretionary spending, potentially affecting R&D budgets and forcing agencies to rationalize their research strategies. The Government Performance and Results Act of 1993 (P.L. 103-62) required all agencies, including research-supporting agencies, to submit measurable goals and performance criteria. In short, for a host of reasons we felt that the time had come to revisit the subject of research priorities, strategies, and contributions to social and economic goals.

Major Themes

Each chapter in this book begins with the implicit or explicit assumption that the current research system is heavily needs-driven and not only curiosity-driven. The relevant concerns are not only how to turn the nation's technical enterprise toward more concrete and practical ends, as the problem is depicted in some of the public debate, but also to judge the contributions of a large and heavily targeted national research investment, to assess the differences in social and private rates of return from research, and to throw light on the most effective institutional roles for the various types of research performers as well as the appropriate mix of public and private support for research. An overall national strategy will almost surely involve some mixture of public and private support for research, but the level and the specific thrust of that support remain at issue. How much to articulate central priorities rather than to follow a more decentralized strategy of setting social goals and linking technical effort to those goals is a related issue.

Harvey Brooks traces the tensions between research autonomy and social direction back to the 1930s debate in Great Britain between Michael Polanyi and J. D. Bernal. Polanyi stressed the need for scientific autonomy and self-governance if research were to contribute most creatively to society, while Bernal foresaw greater need for the large-scale mobilization of research to achieve explicitly formulated social goals. Brooks views the tensions between the two approaches as healthy and unavoidable. Brooks notes that the usual distinction between basic and applied research mixes two separate issues: the matter of top-down versus bottom-up research management and the roles of generalists versus specialists in the choice of research goals. Various combinations are possible that can blend central direction with specialized expertise in the choice of research projects and in efforts to solve social problems. He devotes attention to the educational role of scientists and sees the need for emphasizing greater diffusion of knowledge instead of only knowledge generation in scientific effort and in federal research policies.

Richard R. Nelson and Paul M. Romer do not find cause for alarm in a more explicit orientation of the universities toward the fulfillment of social need. They are uncomfortable with the tendency of some scientists to insist almost as a matter of principle on the nonutility of their research. Nelson and Romer point out that some of the most interesting scientific discoveries and the most significant applications have come from work of a problem-solving character that was neither wholly curiosity-driven nor wholly needs-driven. Echoing a theme from Donald E. Stokes, they argue that scientific work in Pasteur's quadrant is a highly appropriate and common model.[3] Louis Pasteur was a scientist intensely interested in fundamental scientific concepts but whose work was heavily influenced by practical problems arising in medicine and in the industries of his day. Nelson and Romer caution that a strategy pushing too strongly or exclusively for either targeted strategic research or investigator-driven effort would be mistaken.

On the contribution of technical advance to economic growth, Nelson and Romer propose a useful distinction between hardware (all the non-human objects used in production, including capital, goods, structures, land, and raw materials), software (knowledge or information on paper, data on a computer disk, images, drawings, blueprints, and so on), and wetware (the "wet" computer of the human brain, human capital, tacit knowledge, management practices). Their analysis draws attention to the interplay among the three types of inputs and attempts to assess the significance of each. The relative importance of the accumulation of physical and human capital in economic growth is underscored, unlike earlier estimates stressing the importance of technology. In addition, Nelson and Romer propose revisions in intellectual property laws for software as a means to spur innovative activity.

Michael J. Boskin and Lawrence J. Lau also attach importance to improvements in capital stock as a central factor in productivity growth for the G-7 nations in the postwar period. Although their analysis of the sources of U.S. economic growth finds a smaller residual effect directly attributable to R&D than earlier studies, they caution that conventional measures may underestimate the impact of R&D as it interacts with other factors and achieves synergy. Perhaps the most important finding of the Boskin and Lau study is the complementarity between R&D and human capital, human capital and tangible capital, and technology and tangible capital that produces significant interactive effects. Thus, for example, the benefits to the economy of R&D in improved microprocessors will depend on the amount of tangible capital that can make use of the faster microprocessors and on the human capital able to use the computers and the other forms of technology, such as advanced software, that are available and that enhance the capabilities of the improved systems. The current state of the econometric art creates, however, a continual need to refer to industry case studies as a kind of benchmark. Hence, Boskin

3. Stokes (forthcoming).

and Lau are brought full-circle to the conclusion that "R&D is important to economic growth, but just how important is a question economists are not yet fully able to answer."

But Bronwyn H. Hall is perhaps the most skeptical about the current state of knowledge. Arguing that in some instances analysts are unable to specify what would be an adequate test of their propositions, she discusses the refractory methodological difficulties of assessing the impact on industrial innovation and company behavior of university-based and government research. From background research to more focused R&D in the individual firm and then to the marketplace is a long chain of inferences, assumptions, and interpretative leaps. She acknowledges that overwhelming evidence exists that some types of research produce some positive externalities and benefits for society, but she insists on the need for a more satisfying rationale for appropriate levels of R&D investment and on what types of government R&D investment are most productive and necessary.

In contrast, Mansfield presents a decidedly more upbeat appraisal that partly reflects his reliance on industry case studies instead of econometric estimation of production functions. In presenting his paper, he engagingly stated his attachment to an analytical framework "where you can call each data point by its first name." The weight of the empirical evidence is clearly in favor of a high rate of return from research. High rates of return from research are more clearly evident in certain industries than in others, but across the board the social rates of return from research are substantial. Altering assumptions and imposing the most conservative methodological assumptions did not shake his overall conclusions that society benefits heavily from research investments.

Assessing private rates of return is more of a problem because technology diffuses rapidly from firm to firm, sector to sector, and nation to nation. Mansfield devotes a section to explaining how Japanese firms, through heavy investments in process technologies, have been able to make effective and rapid use of research done elsewhere. Whether a firm or country can pursue exclusively or continue to rely mainly on a follower and borrower strategy vis-à-vis technology is more doubtful, however. Investments in technical training, applications, applied research, and testing and equipment seem increasingly necessary for the absorption of technology.

Mansfield attempts to delineate the links between basic research done in universities and technical advance in industry. He finds substantial evidence that basic research has contributed to industrial innovation. Data obtained from seventy-six private companies in seven industries show that about 10 percent of new products and procedures could not have been developed—at least without great delays—in the absence of recent academic research. Factors that encourage the industries to rely on specific academic institutions include the quality and reputation of the faculty, R&D activity in the relevant field, and the convenient geographic location. Although he does not believe that any single pattern of university-industry relationship should apply to all universities, Mansfield advocates closer university-industry linkages in general and sees good reasons for those universities with traditional close ties to certain industries and firms to go farther in this direction. Mansfield identifies some key areas for needed further research. Firms within the same industry often show significant differences in the degree to which they rely on academic research in their new products and processes. Why is that? Considerable reliance also has been placed on some academic researchers as consultants instead of on formal research grants to universities, and this role of the consultant is another promising area for future research.

Most out of step with her colleagues in some respects is Susan E. Cozzens on the subject of quality of life returns from basic research. She frames her topic in unconventional terms by focusing primarily on the process of research and on the values that guide individual researchers, not on research outcomes or traditional institutional relationships. Drawing on biomedicine as a case in point, she studies the issue of the quality of life returns from the pattern of national investments in medical education and research from a one-hundred-year and then from a fifty-year perspective. The Flexner medical reform movement early in the twentieth century produced a more centralized system of training for doctors in fewer institutions of higher quality, but they had the negative effect of leaving rural areas and certain regions of the country without access to local doctors. The gains in health and life expectancy during this period may have been produced more directly by improvements in public health and sanitation measures than by the discovery of vaccines in the new scientific medical schools. Similarly, the great expansion of research support that boosted the National Institutes of Health (NIH) after World War II produced much new knowledge but widened the gap between research and medical practice and did not enrich undergraduate curricula in the universities. Cozzens calls for a process that would incorporate a more explicit awareness and discussion of the broad goals to be accomplished by research in advance of performing the research, and she seeks a form of dialogue between the researchers and the end-users or consumers of the research before new research programs are undertaken. Researchers could then, in her view, internalize a wider set of values and guide their specific research decisions by a more socially aware and relevant metric. She does not explicitly state what criteria government funding agencies might use to evaluate research processes instead of outcomes and whether her approach would strengthen or weaken popular support for science.

The contributions to this volume and the conference discussions only touched on the underlying political climate for the support of research. Some of the attendees at the conference presumed a broad base of support existed for research and wanted to debate the advantages and disadvantages of expanding the university mission to include a more direct contribution to industry and to the economy. The tacit assumption was that universities could enlarge their mission if they were willing to take on the additional burdens of close industry ties but could otherwise simply retain their current roles and levels of research support. Other participants viewed new civilian technology missions as essential to offset inevitable reductions in defense research expeditures. Some doubted, however, whether the defense reductions would have the same impact on universities as on the government's own laboratories and the defense industry. Other participants believed that the push for strategic research that was manifested in the Mikulski proposal could be only the tip of a larger iceberg and that society seemed bound to impose stricter standards of relevance on the scientific community, whatever happened in domestic politics and whatever the level of defense spending. Most participants saw the familiar boundaries between the government laboratories, industry, and the universities as blurring, and thus believed that new opportunities as well as new vulnerabilities could emerge in the post–cold war research system. Government labs, industrial firms, and universities could become interlinked in new cooperative ways but might also find themselves in some ways more competitive than before.

The political earthquake of the 1994 midterm elections underscores the changing circumstances affecting the research system. Republicans were swept into power in the U.S. Senate and House of Representatives, controlled eight of the nine large-state governorships, and achieved parity overall in state legislative seats. The Clinton administration technology strategy to augment the nation's civil technology efforts seemed to be in jeopardy as a result of the election. The Republican House of Representatives had by summer 1995 cut back sharply the administration's budget request for the Defense Department's Technology Reinvestment Program and for the Commerce Department's Advanced Technology Program (ATP). The Senate partially restored some of the cuts.

Republican and Democratic perspectives on R&D have roughly mirrored their party differences generally. Republicans in Congress in general have strongly supported basic research but have been skeptical of federal programs to aid civil technology. They have been less interested in a "dual use" role for defense R&D and seem to prefer a return to a heavy reliance on explicit national security objectives in defense R&D support. Republicans are also more disposed toward cost-benefit and risk assessment approaches in environmental regulation than their Democratic colleagues. Moreover, Republicans, in general, have favored smaller government and tighter controls on discretionary as well as entitlement spending. Space, defense, energy, and health research might be differently affected, but to the extent that government agencies shrink in size, the research supported by the agencies will also shrink. The pressures toward performing more strategic research—as called for by Senator Mikulski—will likely be diminished by the Republican ascendancy.

The Clinton administration and the Congress are stalemated over federal spending priorities in coming years, but both sides have agreed to balance the federal budget by the year 2002, and this consensus will almost certainly result in strong downward pressure on the discretionary elements of the federal budget, most particularly on R&D. While a great deal of attention has focused on the Republican Contract with America, which, over the next seven years, according to calculations of the American Association for the Advancement of Science, would result in a 30 percent reduction in the federal civilian R&D budget, the Clinton administration's agreement also to balance the total federal budget within seven years means that it will be forced to accept reductions of the same magnitude. In the face of these realities, the scientific community seems torn between keeping a low profile or aggressively mobilizing to defend and justify itself.

The dilemma raised by Vannevar Bush in 1945 may yet become a central issue for debate in U.S. science policy.[4] Bush called for a large central research foundation to handle all aspects of federal research support and technology development. He was convinced on the basis of his wartime experience that an operating agency would not support adequately the technical mission and long-term research. Short-run pressures would inevitably force the operating department to cut back on long-term research and technology development in favor of near-term priorities. Bush wanted the whole spectrum of R&D support to remain in the

4. Smith (1990, chapter 3).

central research foundation, up to and including what would now be considered advanced or engineering development. The nation chose, however, to move in a different direction after World War II, relying primarily on the mission agencies themselves for R&D support. Will short-term budgetary pressures now imperil the broad base of support through the mission agencies that has prevailed for so long? Recently, however, proposals for a consolidation of federal science programs in a new department of science and technology have been advanced by the chairman of the House Committee on Science and Technology, Robert S. Walker (R.-Pa.). The issue will certainly continue to be debated in the coming years. If the Republicans retain control of Congress, they are likely to move forward on proposals to consolidate and abolish whole departments of government, resulting in major changes for some mission R&D programs—particularly those associated with the departments of Commerce and Energy.

Even if the social returns from research are demonstrated convincingly in analytic terms, the political process moves according to its own logic, propelled by forces beyond science. The cuts that have afflicted the research systems of other Western nations and the countries of the former Soviet empire in recent years have been dramatic compared with what has happened thus far in the United States. Some conference participants voiced the thought that American science might be fortunate to retain present levels of research support; in this view, the R&D community should seek an orderly contraction of its overall size to a level that can be sustained into the foreseeable future. Scientific "birth control" was not a concept discussed at the conferences, but keen interest was expressed in topics such as addressing scientific priorities, wringing inefficiencies out of the overall system, and possibly promoting partnerships among the various categories of R&D performers.

Future Research

Mansfield's suggestions for future research tend to focus on the activities of academic scientific consultants as a spur to technology transfer and on the interfirm differences in the contributions of recent university research to growth. In the former suggestion, he is not too far removed from some ideas discussed by Cozzens, who also encourages retrospective evaluations of research results as an aid to effective management and the setting of priorities for new research. Nelson and Romer have a number of ideas, but their call for renewed attention to the patent system and to intellectual property rights as a policy concern is among the most important and original of their suggestions. The analytical path trod by Boskin and Lau does not lend itself to ready policy conclusions. Their effort to construct cross-material and time-series data into a new production function for describing the sources of growth, however, is a major achievement and promises to yield further dividends. Similarly, Hall's contribution is of a conceptual nature with potentially broad relevance to practical concerns and public policy issues. She focuses mainly on a more precise identification of the contributions of specific types of R&D investment. Her suggestion for further research into the price deflators used in measuring the returns to R&D is original and important, as is her discussion of the computer industry and how quality change in output induced by R&D can be handled statistically. She also calls for future research on the downstream benefits of government investments in basic science using a variety of bibliometric measures.

Among the topics calling for attention is the impact of recent changes in the organization of industrial R&D. The reorganization and downsizing of central research laboratories in favor of decentralization should be carefully analyzed. The state of industrial research has long been in need of scholarly attention. More needs to be known about how and how much R&D is conducted in industry, in what kind and size of firm, and whether the United States neglects process technologies in comparison with some other nations.[5] What, furthermore, are the implications of the new strategic alliances that are energizing many multinational business firms? Is collaborative research activity inherently defensive and technologically conservative in nature? Or, alternatively, is collaborative international effort the scale of commitment needed for major innovations and thus the wave of the future?

Numerous policy questions remain for government research funding agencies, university administrators, and corporate strategists and technologists. The need to retain a high technical competence across the broad frontiers of science and technology is a national priority according to nearly all of the conferees, but behind this formula many important policy questions remain relating to the size, the breadth, and the focus of government research investment. To what extent should American universities become international training centers, and how can they retain a culture of open science if they become more closely involved with proprietary research for either domestic or foreign companies? How far

5. Fusfeld (1994).

should government laboratories go in international collaborative activities with governmental technical institutes from other countries?

Brooks singles out a cluster of unresolved issues relating to the linkage between technology and economic performance. If job creation is a central concern, he queries whether the current large centers of technological capacity are best suited to achieve the desired end. The national laboratories, he suggests, might need to reorganize into smaller units that more closely match economic needs (even while retaining geographical proximity for various purposes). Similarly, relatively little research has focused on the effectiveness and the impact of the Cooperative Research and Development Agreements (CRADA) now widely pursued by federal agencies to foster innovative activity by transferring government technologies.[6] If large business enterprises suffer from high-cost overhead structures, have the large universities also grown too big and centralized? Whether government R&D funds merely displace private investment or actually augment it is another traditional issue that still invites analysis.

The questions that this book set out to answer were highly challenging, and over the years many of the nation's leading economists, scientists, and science policy analysts have wrestled with them. Many important questions remain unanswered, and changing circumstances have generated a host of new analytical and policy concerns. We hope that the present volume will spur further efforts by colleagues from a variety of disciplines to analyze the cluster of significant issues relating to the social and economic returns from research.

References

Bozeman, Barry, and Michael Crow. 1995. "Federal Laboratories in the National Innovative System." Georgia Institute of Technoloy, School of Public Policy.

Denison, Edward. 1967. *Why Growth Rates Differ: Postwar Exerperience in Nine Western Countries*. Brookings.

———. 1979. *Accounting for Slower Economic Growth*. Brookings.

———. 1985. *Trends in American Economic Growth*. Brookings.

Fusfeld, Herbert I. 1994. *Industry's Future: Changing Patterns of Industrial Research*. Washington: American Chemical Society.

Smith, Bruce L. R. 1990. *American Science Policy since World War II*. Brookings.

Solow, Robert. 1957. "Technical Change and the Aggregate Production Function." *Review of Economics and Statistics* 39: 312–20.

Stokes, Donald E. Forthcoming. *Pasteur's Quadrant: Basic Science and Technological Innovation*. Brookings.

6. See, however, Bozeman and Crow (1995).

CHAPTER 2

The Evolution of U.S. Science Policy

Harvey Brooks

T HE CURRENT national debate over science policy can trace its lineage to an argument that took place in Great Britain in the 1930s between Michael Polanyi and J. D. Bernal. Then, as now, the debate concerned the degree to which planning the agenda of the national science and technology enterprise to achieve explicit social or economic goals is feasible and desirable. Polanyi stressed the need for autonomy and self-governance of the scientific community if it were to contribute most efficiently to societal goals in the long run. His view was succinctly summarized by the sociologist Bernard Barber: "However much pure science may eventually be applied to some other social purpose than the construction of conceptual schemes for their own sake, its autonomy in whatever run of time is required for this latter purpose is the essential condition of any long run applied effects it may have."[1] In contrast, Bernal, who was strongly influenced by Marxist thought, saw tremendous inefficiencies with autonomous science and believed that its enormous potential benefits for humanity could be realized only through a publicly debated plan involving government and many representative elements of society.[2] In 1934 Bernal estimated that the United Kingdom was spending only 0.1 percent of its gross national product (GNP) on research and development (R&D); the United States, 0.6 percent; and the USSR, 0.8 percent. In 1939 he called for a tenfold increase in Britain,

1. Barber (1962, p. 139).
2. An excellent summary of the Polanyi-Bernal debate of the 1930s is given in chapter 1 of Freeman (1993). The quote is from p. 21. According to Freeman, Bernal felt a gentlemen's agreement existed to gloss over the internal inefficiencies of science for fear of losing even the present inadequate resources that science gets.

a level that was achieved twenty-five years later.[3] Bernal saw the USSR as a hopeful model of the potential contribution of science to society, based partly on his implicit acceptance of Marxism as a social science rather than as a political ideology. Later he came to share the general disappointment with the social and economic returns from R&D in the socialist countries that made the changes that he had suggested.[4] But he may have also recognized that this relatively low payoff from increased R&D was partly the result of the concentration on military-oriented R&D and that the spinoff of a few high-tech industries from military R&D could not offset the lack of R&D investment in a much wider range of industries.[5]

The Bush versus Kilgore Debate

At the close of World War II, many features of the Polanyi-Bernal debate of the 1930s were renewed in the United States over competing visions of the future of American science policy. The wartime experience helped to shape the debate. Vannevar Bush became the exponent of the Polanyi view, articulated in the famous 1945 report *Science: The Endless Frontier,* while Senator Harley Martin Kilgore, D-W.Va., became the exponent of the Bernal view, with more active government direction of the research agenda.

In his recommendations to President Harry S. Truman in 1945, Bush selected the university as the centerpiece of postwar science policy specifically because of its independence and autonomy:

> It is chiefly in these institutions [universities] that scientists may work in an atmosphere which is relatively free from the adverse pressure of convention, prejudice or commercial necessity. At their best they provide a substantial degree of personal intellectual freedom. . . .
>
> Satisfactory progress in basic science seldom occurs under conditions prevailing in the normal industrial laboratory. There are some notable exceptions, it is true, but even in such cases it is rarely

3. Freeman (1993, pp. 7, 10); Bernal (1978, pp. 62, 242).
4. Freeman (1993, p. 14); Bernal (1958).
5. Freeman (1993, pp. 15–16); see also Alic and others (1992), especially chapter 12.

possible to match the universities in respect to the freedom which is so important to scientific discovery.[6]

Partly because of the unforeseen advent of the cold war before the establishment of the National Science Foundation (NSF), the universities came to represent only a fraction of public spending on R&D (about 9 percent, rising to 12 percent for universities proper and eventually to about 16 percent, including separately organized, government-owned R&D centers operated under contract by universities). Nevertheless, universities remained central to the debate about American science policy. They employed the majority of Ph.D.s in science and in engineering, performed the majority of what was defined as basic research sponsored by the federal government, supplied a majority of the science advisers to government agencies, and were looked to by government as a first option when the need for new initiatives and institutional arrangements relating to science and technology were identified by policymakers.[7] The average person who follows the press and other media is frequently incredulous when told what a small proportion of federally funded R&D is performed in universities (in the mid-1990s, for example, only about half that performed in government-owned laboratories). According to some scholars, Bush's statement quoted above served its author's purposes in urging expanded federal support for the conduct of basic research by universities, and it reflects the basis of the current justifications for public patronage of university-based research in the domain of science and technology.[8]

Senator Kilgore was equally bullish regarding the support of science and its promise for the achievement of social and economic goals, but he called for a much closer linkage to political institutions and less autonomy for the scientific community in setting the research agenda. The model for science envisioned by Kilgore was akin to that established in the nineteenth century for the agricultural research system, the extension service, and the land grant universities based in the states, with more emphasis on the wide geographical distribution of research capacity,

6. (Bush and others, 1960, p. 19). Bush's observation seems to be supported by the gradual migration to academia of some of the most creative and productive scientists even from such exceptional industrial laboratories as the Bell Laboratories, the General Electric Research Laboratory, the various IBM Corporate R&D centers, and several other notable industrial laboratories with a long history and strong traditions.

7. Brooks (1986a).

8. David, Mowery, and Steinmueller (1994).

stricter political accountability, the diffusion of new knowledge, and its adaptation to the needs of potential users.

Nor was the scientific community unanimously on the side of Bush. Important and prestigious scientists such as E. U. Condon and Harlow Shapley favored the Kilgore view. For the most part, however, the outcome of the debate was a fairly conclusive victory for the Bush approach, with the important exception that the director of the new National Science Foundation was to be appointed by the president and confirmed by the Senate, not by the National Science Board without Senate confirmation as originally proposed. The controversy over political accountability and responsiveness of the NSF delayed its establishment for nearly five years, allowing time for the habits and the culture of the support system to evolve as an adaptation of the military R&D system inherited from the war effort.[9]

The Watershed of World War II

The United States is unique among the industrialized countries in how its R&D system and science policy was transformed permanently by World War II. In both Britain and America, science and technology were mobilized on an unprecedented scale—more or less the scale that Bernal had in mind, only for war and not social and economic development as he had advocated. In the United States, R&D as a proportion of a much larger gross domestic product (GDP) reached about 0.8 percent in 1945.[10] Even with respect to the military, the role of science and scientists was far different in World War II from what it had been in World War I or in the interwar years. Then, military research was performed largely in civil service laboratories under direct military supervision, and weapons were purchased from government arsenals or from industry to detailed specifications set by the military with relatively little support from R&D. In the case of industry, R&D consisted mostly of experimental design and testing, which were integral parts of the procurement process. The separate R&D contract for development of a weapons system was largely unknown, as was the simultaneous parallel pursuit

9. See, for example, Sapolsky (1990).
10. According to Galbraith, gross domestic product (GDP) increased by 80 percent between 1939 and 1944. See Galbraith (1994, p. 115). *McGraw-Hill Yearbook of Science and Technology* (1963, pp. 11–21; table 1, p. 13).

through R&D of alternative approaches to the design of a given weapons system.

In 1935 the U.S. federal government accounted for only 13 percent of total national expenditures for R&D, which amounted to only 0.35 percent of national income; the federal contribution to R&D was thus only 0.05 percent of national income. As late as 1938, agricultural research was about 40 percent of federal R&D (more than military R&D, which accounted for only about 25 percent).[11] Total public R&D represented only 0.25 percent of the federal budget in the 1930s. Both agricultural and military R&D were performed principally in civil service laboratories and supported primarily on a level-of-effort basis institutionally rather than a project-by-project basis, so that expanding effort rapidly was difficult as unforeseen possibilities were revealed by research.

By 1962 the federal share of total national R&D support had risen to nearly 70 percent and represented 11 percent of the discretionary spending in the federal budget. Of federal R&D expenditures in 1962, nearly 93 percent were in defense, space, and atomic energy—all essentially deriving from the cold war. Agricultural research, though larger in real dollars than in 1938, was now only 1.6 percent of federal R&D. As to the actual performance of research (as opposed to federal support flowing to external research-performing entities), only 14 percent was performed in laboratories staffed by civil servants. An additional 6 percent was in wholly owned government facilities run by contractor personnel (known as Federally Funded Research and Development Centers, or FFRDCs). The balance of 80 percent was contracted out to private industry, independent private not-for-profit laboratories, and colleges or universities.[12]

As an indication of the climate of opinion at the time, many in the military on the eve of World War II saw increased investment in R&D as delaying the procurement of badly needed weapons of existing design.[13] Largely on the initiative of the university scientific community, led by Vannevar Bush and James B. Conant, American scientists and engineers began to mobilize for the impending war effort. Particularly involved were the nuclear physicists, who were at the forefront of the

11. Brooks (1968, p. 24).

12. These statistics are taken from a variety of sources. They are summarized in Brooks (1990, pp. 11–12).

13. According to an Army General Staff report issued at the time, "The Army needs large quantities of excellent equipment that has already been developed" but "the amount of funds allocated to research and development in former years is in excess of the proportion for that item in consideration of the rearmament program." See Dupree (1957, p. 367).

most exciting science of the times. They began to mobilize in a few university-based laboratories, where they could work with their own colleagues in a familiar environment. Methods of contracting for R&D at these university-managed laboratories were rapidly devised, based on the principle of full-cost reimbursement (including indirect costs, an accounting innovation that has continued to be unique to the United States for government support of research) and "no gain, no loss" financially to the institutions or individuals who joined the effort. The laboratories, in turn, undertook to provide their best efforts with no guarantees as to outcome. As the United States entered the war and urgency increased, the simultaneous pursuit of parallel technical paths to solve military problems became standard practice in the belief that the extra costs of possibly duplicative R&D could be ignored, given the losses that might be incurred by military failure.[14] This "cost is no constraint" approach was facilitated by the technical resources of the nation, like its other productive resources, which were grossly underused in 1939.[15] The parallel project approach assumed its most extreme form in the Manhattan Project for the production of the atomic bomb.

The most dramatic technical developments of World War II emerged from the university-centered wartime laboratories organized under the auspices of the Office of Scientific Research and Development (OSRD), a civilian agency reporting directly to President Franklin D. Roosevelt and headed by Vannevar Bush. Bush had the authority to undertake strategic research initiatives independently of the War Department and the military services, although in practice he usually cooperated closely with them. To Bush, an engineer and inventor and not a scientist, the arrangement was a lesson in the necessity for autonomy and independence in technical decisionmaking. The most dramatic and visible parts of the technical agenda were in the hands of people whose backgrounds and original acculturation had been in the world of academic science, particularly the international fraternity of nuclear physics, where engi-

14. This came to be known as the "whole problem approach." In the words of one observer, scientists "were eventually able to persuade the soldiers to inform them of the general military problems involved in order that the scientists might reach their own conclusions about the kinds of weapons and devices the military would need to meet these problems." See Schilling (1964).

15. Unemployment in the United States was at 17.1 percent in 1939 compared with 3.9 percent in 1947. See Galbraith (1994, p. 136). By contrast, the German economy was already at close to full employment at the beginning of the war.

neering skills in the design of complicated apparatus were at a premium. The physicists had already developed a tradition of teamwork, which was still absent from most of the rest of academic science. The social organization of the nuclear physics community at that time was uniquely adapted to meeting the particular technical challenges of World War II and helped set the pattern of thinking that heavily influenced the evolution of science policy during the cold war and beyond.

Among other things, this pattern of thinking tended to reinforce a linear model of innovation in which research largely preceded engineering and development. Research thus dominated the innovation agenda even though it absorbed only a small fraction of the total resources and human effort going into the innovation process. Descriptions of the innovation process emphasized the flow of information from science to engineering, and then successively to production and to the market. Overlooked were the reverse flows back along the chain from the market to production and to research as well as the leaps that often bypassed the intermediate steps in the linear description. Thus, in the idealized linear model, innovation begins with basic research that turns up discoveries while being pursued almost without thought of application. These discoveries in turn suggest opportunities for applications that are pursued through applied research, development, design, production, and marketing. In this model, the rest of the innovation chain cannot exist without basic research, which is the foundation on which the productivity of all subsequent investments depends. This simplistic model, though increasingly challenged by scholarly research, has had an important and persistent influence on the organization and management of innovation in the United States until recently. It still provides a plausible description for many of the radical paradigm-shifting technological innovations during the cold war period, not only in defense but also in a number of commercial areas.[16] The most classic case is the development of nuclear weapons and nuclear power following the discovery and theoretical explanation of nuclear fission, but other examples include the transistor, the laser, genetic engineering, and many biomedical technologies.[17] None of these

16. For an analysis of the types of innovations for which the linear sequential or staged model is reasonably appropriate, see Lee (1992).

17. For some recent examples in the mathematical and physical sciences, see National Science Foundation (1994), especially "Origins of the Information Superhighway: A Retrospective Look at the Discoveries That Made Possible the Emerging Technologies of Today and the Services of Tomorrow," pp. 12–16.

cases is as pure as the linear model predicted, however, and their success depended heavily on follow-on investments for incremental improvement. Nevertheless, enough truth exists to suggest that abandoning the linear model would be a serious mistake.

The linear model works best for a firm when: (1) the company is seeking radical product innovations that do not fit into its existing business and marketing organization; (2) the new product is expected to have a long shelf life without the necessity for many new models or fine product differentiation; (3) the technical resources of the company and its R&D investment are large compared with those of its potential competitors; (4) the market is protected by oligopoly, a legal monopoly, or public procurement policies biased in favor of domestic suppliers; (5) the nature of the demand is such that the trade-off between cost and performance tends to be weighted heavily in favor of performance; and (6) the total number of units sold is expected to be relatively small. The bias toward performance over cost tends to be particularly marked in the case of defense applications and curative biomedical technologies but is also true for major categories of capital equipment. All six of the criteria were present in much of the weapons industry, the nuclear power industry, the telecommunications industry, and, in some degree at least, the U.S. computer industry in the early period of the cold war, when the biggest market was still the military or the space program. In some measure the isolation of the R&D function was an advantage, because new ideas were less constrained by existing organizational routines and concepts of market demand. For novel products, customers usually do not know what they want until they have some notion of what they can get—in other words, of what is technically possible. The history of radical innovations is studded with ludicrous underestimates of potential markets, such as the prediction attributed to Thomas Watson that the world market for computers might eventually be five large computers. The lack of interest of many large companies such as Kodak in the early concepts for xerography is another illustration of the same phenomenon.

However, the persistence of the linear mental model of innovation probably led to an organizational trend in U.S. industry that was ill adapted to the emerging international competitive environment in the 1970s and 1980s. Many large companies had established corporate R&D centers on idyllic rural campuses far removed from their business divisions in the expectation that this was the best way to produce breakthrough innovations that would replace the mature businesses that some corporations seemed prepared to abandon to low-wage countries, according to the product cycle theory fashionable in the 1960s.[18] This mindset proved especially dysfunctional for the U.S. automobile industry.[19]

The Challenge of the 1970s

The years 1967 and 1968 marked the high-water mark of the post–World War II federal investment in R&D in the United States and a turning point in U.S. science policy. Up until then, science policy had been dominated by the cold war. By 1963 the national investment in R&D was approaching 3 percent of GDP, 2.5 times the peak reached just before the end of World War II, and more than 70 percent of this effort was supported by the federal government. Of this, 93 percent came from only three federal agencies: the Department of Defense (DOD), the Atomic Energy Commission (AEC), and the National Aeronautics and Space Administration (NASA)—all largely driven by the rivalry with the Soviet Union. The technology management styles of these three agencies were similar, although the organic statutes of both AEC and NASA gave them a more explicit mission of ensuring the commercial spinoff of the knowledge and technologies they had developed. Much of the rest of federal R&D was in the biomedical field; the budget of the National Institutes of Health (NIH) had soared since about 1956 and was growing at close to 30 percent a year. During this period, 55 percent of the federally financed research performed in universities was in the life sciences, supported mostly by NIH and secondarily by the Department of Agriculture.

Federal support for R&D as a whole and for (mainly) basic research in universities reached a peak in 1967, after which it declined in real terms until about 1976. A partial exception was the life sciences, which continued to grow but at a much diminished rate. In the mathematical and physical sciences and engineering, a drop occurred in actual or nominal dollars; the drop was about 14 percent below the 1967 peak in real terms.[20] It was a period of sharp decline in job opportunities for scientists and engineers, especially junior-level faculty appointments in universities. Unemployed scientists and engineers received considerable attention in

18. Vernon (1966).
19. Florida and Kenney (1990); see also Dertouzos, Lester, and Solow (1992).
20. Brooks (1986b, pp. 119–67).

the media and in Congress, even though their unemployment rate remained well below that of skilled production workers and technicians.[21]

More germane here were the challenges to the linear model of innovation. An amendment sponsored by Senator Mike Mansfield, D-Mont., to the fiscal 1970 defense authorization bill required that "none of the funds authorized to be appropriated by this Act may be used to carry out any research project or study unless such project or study has a direct or apparent relationship to a specific military function or operation."[22] The Mansfield amendment was directed against general purpose basic research funded by DOD. Although the legal force of the amendment applied for only one budget year, it sent a strong message to all research funding agencies and resulted in the transfer of several long-term DOD university research programs to civilian agencies, usually with diminished funding. For example, the innovative Materials Research Laboratories started by the Advanced Research Projects Agency (ARPA) of DOD in 1961 were moved to the National Science Foundation. The National Magnet Laboratory at the Massachusetts Institute of Technology (MIT), started by the Air Force, was also transferred to NSF sponsorship. Several DOD-sponsored programs involving the social sciences were canceled.

Another challenge came from a widely publicized DOD-sponsored study known as Project Hindsight, which was widely interpreted as showing that few ideas originating from basic research had contributed to specific DOD weapons.[23] This argument was seized upon by critics to denigrate the value of federal sponsorship of academic research in general. A counterstudy commissioned by NSF called Technology in Retrospect and Critical Events in Science (TRACES) sought to show that the Hindsight study had utilized too short a time horizon in trying to identify the basic research events that had contributed to technological advances. However, the TRACES study dealt with wholly different examples, mostly from the commercial area, so that directly comparing the two sets of results was difficult.[24]

News commentators, citizens, college students, and members of Con-

---

21. Brooks (1977).

22. P.L. 91-121, Section 203. For Senator Mike Mansfield's explanation of the rationale behind this amendment, see U.S. Senate (1970, pp. 604–09). See also Smith (1990, pp. 78–85).

23. Sherwin and others (1966); Isenson (1969).

24. Illinois Institute of Technology (1968). Mowery and Rosenberg show that the concept of research events is much too simplistic and that neither Hindsight nor Technology in Retrospect and Critical Events in Science (TRACES) employs a methodology that is capable of showing what they purport to show. See Mowery and Rosenberg (1982).

gress were also pressuring the scientific community in many forums to turn away from curiosity-driven basic research to pressing social problems, especially those that had been raised to high public visibility by the Great Society programs of President Lyndon B. Johnson. A considerable turning away from science by college undergraduates became evident, while at the graduate and postdoctoral level interest in the social sciences expanded.[25] Some claims advanced on behalf of the social sciences at this time seem overblown and simplistic in retrospect, often amounting to the advocacy of social engineering with strong manipulative overtones. This eventually helped to trigger a backlash against federal support of the social sciences when the Reagan administration assumed power in the 1980s. Nevertheless, this period did produce some ultimately useful soul-searching on the part of the entire scientific community.

One result of the pressures on NSF was the invention of a new program known as Research Applied to National Needs (RANN). A separate part of NSF, RANN consisted of a series of interdisciplinary projects carried out by teams of scholars in universities drawn from several different disciplines and focused on a particular societal problem. It received mixed reviews, especially for those projects involving a large social science component.[26] The most successful programs were those dealing with energy conservation, renewable resources, certain environmental problems, and solar energy. These served to some extent as pilot programs for the much larger programs that the newly created Department of Energy (DOE) developed in the 1980s.

A parallel program managed by the National Bureau of Standards (NBS)—Experimental Technology Incentives Program (ETIP)—worked mainly with regulatory agencies in designing strategies of regulation that would encourage technological innovation (for example, general performance standards for household appliances in place of restrictive detailed design standards). The program was successful in achieving its limited objectives, but the scale was too small to have any permanent or widespread influence on regulatory strategies as a whole. In the end, both RANN and ETIP failed to take root as continuing programs in their respective sponsoring agencies.

From 1968 to 1975, scholars and a few government researchers began

25. For a contemporary evaluation of the status and future prospects of the social science disciplines, see the Social Science Research Council (1969), the so-called BASS report.

26. National Science Foundation (1978), especially "RANN Programs in the Social and Behavioral Sciences," pp. 63–72.

sounding the alarm about an impending loss of U.S. international competitiveness as a result of an apparent decline in the rate of growth of manufacturing productivity, usually attributed to underinvestment in R&D.[27] The principal response of the Nixon administration was the planning of a large-scale program of federal technological investment, following the model of spinoff from space and defense expenditures that had held sway at DOD, NASA, and AEC.[28] This program never got far because of budgetary stringency.

Another casualty of the political turbulence of this period was the dismantling of the White House science advisory apparatus. The functions of the science advisor and the President's Science Advisory Committee reverted to the director of NSF and the National Science Board (to whom they had been assigned by the law establishing NSF, but who had done little in practice). The functions of the Office of Science and Technology and some of those of the interagency Federal Council on Science and Technology were absorbed by a somewhat augmented NSF staff in the office of the NSF director. This development was viewed with some dismay by the scientific community, but the influence of the original apparatus had already been greatly diminished before the Nixon reorganization.

Shortly after the inauguration of President Gerald R. Ford, interest in the reestablishment of the science advisory apparatus in the White House arose in a number of quarters. Vice President Nelson Rockefeller was asked by President Ford to make recommendations in this regard. In the meanwhile, the National Academy of Sciences also initiated with private funding a comprehensive study of the top-level science organization of the federal government. This study was chaired by James Killian, former president of MIT, who had been President Dwight D. Eisenhower's first science advisor.[29] This combination of initiatives ultimately led to the creation by Congress of a new Office of Science and Technology Policy in the Executive Office of the President and to the reestablishment of the science advisor position, along with some additional associate director

27. Boretsky (1973). For a very early comprehensive academic analysis of this problem, see Hollomon and Harger (1971). See also discussion of Boretsky's gadfly role in Branscomb (1993, p. 65).

28. For a well-documented skeptical study of the technology transfer program in the National Aeronautics and Space Administration (NASA) and the spinoff model in general, see Doctors (1971).

29. National Academy of Sciences (1974); see also Brooks and Skolnikoff (1975) and Smith (1992, chapter 8).

positions. The new apparatus was nominally more influential than the old, though in practice the influence of the science apparatus depends much more on the interests and receptivity of the president than on its formal status. The main strength of the new legislation was that for the first time the science apparatus in the Executive Office had full statutory sanction, so that it became difficult to abolish completely and easier to revive when a more interested president came into power.[30] Previously, the office had been established and disestablished through the president's reorganization authority without explicit legislative sanction.[31]

Resurgence of the Cold War

After the era of social priorities and the collapse of the Nixon administration, there began in 1976, unforeseen by much of the scientific community, a resurgence in federal support of R&D as well as private R&D investment.[32] Much of this was stimulated by the reemergence of foreign affairs at the center of national attention. The Iran hostage crisis, the Soviet invasion of Afghanistan, and the general reintensification of the cold war produced a new military buildup beginning in the second half of the Carter administration and greatly accelerating in the Reagan administration. Two factors were especially important in the increased R&D spending. First, energy policy and, to a lesser extent, health and environmental policy emerged on the national political agenda. The AEC eventually became the Department of Energy, with a much broader charter covering all energy sources, not just nuclear power, and with renewable energy briefly achieving a larger budget than nuclear energy during the Carter administration. Federally supported energy R&D increased by a factor of 4.6 in real terms between 1969 and 1981 but sank back to only 1.9 times its 1969 level by 1994. A parallel increase was seen in private energy R&D investment during the same period. By contrast, defense and space together had fallen to only 58 percent of its 1969 level by 1976 (65 percent of total federal R&D).[33] The falloff in total

30. National Science and Technology Policy, Organization, and Priorities Act of 1976, U.S. Code 6683; see also Branscomb (1993, chapter 1, especially pp. 7–9).
31. Branscomb (1993, p. 7) and Brooks (1986a).
32. I speculated about such a development in the Donald Hamilton Memorial Lecture given at Princeton in 1973. See Brooks (1974).
33. Brooks (1986b, p. 131).

federal R&D after 1967–68 lasted to 1976, when it began to recover slowly.[34]

Second, public concern grew steadily about the loss of U.S. competitiveness in international markets, which was still at that time largely attributed to lack of investment in R&D, particularly civilian R&D. Although the total U.S. R&D investment as a fraction of GDP was comparable to that of Europe and Japan, a much larger fraction was in the defense and space sectors. Economists were increasingly arguing that military R&D produced much less benefit to the economy as a whole per dollar than commercially oriented R&D. Thus a complete reversal occurred of the previous terms of debate in the late 1960s, when J. J. Servan-Schreiber, a French journalist, had written a best-selling book arguing that the U.S. investment in space and defense R&D was inevitably generating an insuperable competitive advantage for the United States in world markets—the so-called technology gap in favor of the Americans.[35]

Already in the early 1970s the argument was turning around to say that the U.S. defense burden was creating a technology gap in the opposite direction, favoring competitors whose investments in space and defense were smaller. The proponents of this view tended to dismiss as irrelevant that civilian R&D and privately financed commercial R&D in the United States far exceeded that of its principal competitors—about the same as that of West Germany, France, the United Kingdom, and Japan combined (as of 1979).[36] This dismissal assumed that virtually no economies of scale exist in R&D and that any spinoff of R&D between different institutions in a country stops abruptly at national boundaries. Nevertheless, the R&D-to-GNP ratio has become a shibboleth of technology and competitiveness arguments that persists to this day.[37] Such arguments apparently had a considerable influence on both private and public decisionmaking, as evidenced by the accelerated growth of private R&D spending after 1976, even through the recession of 1980–82, when many companies were laying off production workers while continuing to expand their technical work forces.[38]

When the Carter administration took office in 1977, one of its earliest

---

34. Brooks (1986b, p. 127).
35. For a later review and critique of Servan-Schreiber's argument, see Brooks (1973, pp. 1, 8–11).
36. Brooks (1986b, p. 136).
37. For an extended discussion of this issue, see Brooks (1985).
38. Brooks (1986b, p. 127).

initiatives was to launch a major study of the impact of federal policies, including R&D, on U.S. competitiveness. The study was to make recommendations for changes in federal policy that would enhance private incentives for innovation and investment in R&D.[39] As the first major official study to emphasize the demand side of R&D policy, it represented a major departure from basing policy on a linear model of technological innovation. Its impact was blunted, however, by a number of foreign policy events that temporarily diverted both public attention and that of decisionmakers from the economic competitiveness problem. The Iran hostage crisis and the second oil embargo were notable, the latter leading to the rapid buildup of energy research and development and demonstration programs. Because of the unprecedented inflation associated with the energy crisis, the underlying (mainly microeconomic) sources of declining U.S. competitiveness were put on the back burner until well into the 1980s, after inflation had been brought under control.

The rapid defense buildup in the Reagan administration peaked in 1985 but probably helped to provide a Keynesian boost to the economy. The defense R&D program of sophisticated weapons such as the strategic defense initiative (SDI) also had the effect of prolonging implicit acceptance of the linear model of innovation until the leveling off of defense spending once again began to expose the deeper structural problems of the civilian economy. Although the Reagan defense buildup was more development- and less research-intensive than the 1960s buildup, basic and academic research still fared well financially in the 1980s, assisted by the continuing boom in biomedical research and the emergence of new commercial opportunities in biotechnology, microelectronics, computers, and optoelectronics. Many of these commercial opportunities were closely linked to basic and academic research, even though the technological developments required larger downstream investments by industry, many of which were venture capital startups.

At the same time, subtle but important changes were taking place in the role of universities in the U.S. research system. One of the most important had to do with intellectual property. Until the mid-1970s it was almost a matter of pride with most university researchers that the academic research system was completely open. Patents were seen as obstructing the free flow of information. The discoveries in molecular biology in the early 1970s for the first time revealed the possibility of

---

39. U.S. Department of Commerce (1979); see also U.S. House of Representatives (1980, especially pp. 155–73).

commercial application of fundamental phenomena hitherto thought to be of only academic interest. Before the late 1970s, if patents were taken out on university research supported by the federal agencies, they were most frequently assigned to the government, which had no particular mechanisms or policies for fostering pursuit of their further development other than for government end-use. No governmentwide patent policy was in place, and each sponsoring agency had its own guidelines ranging from compulsory ownership by government (AEC, NIH, and NASA) to some system of rights for the individual inventor (DoD and Agriculture). Permission to license for commercial exploitation did exist (though exclusive licenses were rare). By the end of the 1980s the situation had dramatically changed. Most universities with significant research programs had established policy guidelines and a special office to deal with intellectual property issues. These offices were becoming more aggressive in patenting inventions made by university personnel and in seeking to license such patents for commercial development to interested companies or to assist the inventor(s) to establish a new company. Legislation enacted in 1980 and 1984 encouraged this process and specifically permitted university ownership of patents on inventions made with the support of government grants or contracts. The legislation also allowed the (possibly exclusive) licensing of these patents to small firms, and subsequently to larger firms, always with a free license for government end-use.[40]

During the 1980s a very dramatic growth also was seen in the number of University-Industry Research Centers (UIRCs) supported jointly by one or more federal agencies, industries, and the states. By 1990 more than one thousand such centers had been identified having an aggregate budget from all sources of more than $4.12 billion, with about 31 percent of the funds on average coming from industry—about three to four times the share of other industry support for academic research. About 60 percent of these centers had been founded after 1980. Although the majority of them saw their role as "principally providing windows on new technological developments" to industry instead of "more direct contributions," 26 percent of the UIRCs surveyed reported "transfer of technology to industry" and "improving industry's products and processes" as significant goals. These observations are consistent with the fact that only 16 percent of their R&D activity was reported as "devel-

---

40. Sandelin (1994); cf. also Public Law 96-515, December 12, 1980, "Patents and Trademark Laws, Amendment," and Public Law 98-622, "The Patent Law Improvement Act," November 8, 1984; and Smith (1990, pp. 58–62).

opment." Relations with industry were spread over a wide range of fields. Furthermore, the UIRCs differed greatly among themselves in the relative amount of support from various sources and in the closeness with which they worked with industry.[41] In the 1960s more than 70 percent of academic research support came from federal agencies; the proportion had dropped to only 57 percent by the 1990s and was still declining. Large differences existed among individual universities, with many private universities being more dependent on federal funds than the state universities.

Economic Performance and Government-Sponsored Research

During the resurgence of the cold war, the pressure for societal direction of research, especially university research, had lost some of its political impetus. The steadily growing resources enjoyed in aggregate by R&D after 1976 allowed a relatively autonomous science culture to live more easily side-by-side with a more societally driven agenda within the universities. But the end of the cold war brought a new surge of interest in tight social direction of science. This time, instead of reflecting goals of the Great Society, interest focused on international economic competition and private sector job creation. For the first time those on the lower rungs of the economic ladder were not the only ones being hurt by the weak performance of the U.S. economy. The middle class was also affected by the restructuring of the economy. Job losses and downsizing of firms resulted as the deregulation of protected monopolies or semiprotected oligopolies took effect. The universities came under increasing pressure to do something about the economy, perhaps stimulated by some of the rhetoric that had often been used in justifying federal support of academic research in the past.

A subtheme of the debate was the tension between open science, represented institutionally by the elite research universities, and "appropriable science," represented by influential segments of industry and politicians in support of "technomercantilism."[42] Many paradoxes and crosscurrents arose in the debate not congruent with political ideologies

41. Cohen, Florida, and Goe (1994). On average about 34 percent of the support for these centers was provided by federal agencies, a far smaller fraction than for university research overall. About 12 percent came from state funds and 18 percent from internal university funds.
42. Dasgupta and David (1992).

in other areas. A second subtheme related to the degree to which social need rather than scientific opportunity should govern research priorities, especially for academic research. The sharp dichotomy between the Polanyi view of autonomous science and the Bernal view of science entirely driven by social need seems increasingly artificial. Most science and technology are, and must be, driven by social need, so the real issue is the proportion of the total public resources for research that should be driven primarily by the internal logic of the subject (scientific opportunity), not an either/or choice between extremes that are seldom seen in their pure form. Further complicating the issue is that in practice much significant fundamental science has been derived from practical problems first encountered in applications but subsequently pursued much more deeply in a conceptually oriented mode. Thus the real choice lies in how far to encourage investigators to go beyond the immediate need for a solution to a practical problem close at hand to the more general conceptual problems that arise in the course of the research. Often entirely new domains of fundamental research open up that would not otherwise have been identified in the absence of the original applied problem.[43]

Two separate issues arise in the debate about the allocation of resources for research, which have frequently been confused in public and congressional discussion. One is the proper balance between top-down management by the government sponsors of research—usually a program officer in a federal agency—and reliance on bottom-up generation from the grass roots of the scientific and technical community. The other is the appropriate level and mode of participation of nonscientists and politically accountable generalist policymakers in the design and evaluation of research programs.

A program could be tightly managed from Washington by purely technical experts determining the research agenda of projects in considerable detail, leaving only narrow discretion over strategy to the performer of the research. This tended to be the style in which the NSF RANN program of the early 1970s was managed. Or a general field of research could be determined at a high level in Washington, but scientific peer evaluation and selection of competitive proposals from the grass roots of the scientific community could be depended upon to respond to a broad RFP (Request for Proposal) for the research strategy within the broad program guidelines.

Conversely, an open competition could be held among proposals,

43. Brooks (1980)

originating entirely bottom-up from the scientific community, uncon-strained by need-driven guidelines but utilizing a competitive evaluation process that incorporates the views of laymen and generalists as well as technical experts. Various balances of representation between laymen and experts could be decided at the start, but the societal need component could be allowed to develop directly out of the evaluation process instead of being determined in advance.

Many different combinations of the two dimensions—top-down ver-sus bottom-up and generalist/laymen versus experts—are possible. In the Advanced Technology Program (ATP) of the National Institute of Stan-dards and Technology (NIST), for example, some projects are responsive to broadly defined RFPs and are evaluated primarily by experts, with a certain proportion of the available funds reserved for an unrestricted competition in which the evaluations are made by a mix of experts and generalists. A further complication is the possibility of a staged project evaluation process in which proposals are first screened by experts for technical quality, and then the technically high-quality projects are further winnowed by more broadly constituted panels of experts and laymen.

If the process of using science for social purposes is thought of as one of optimally matching scientific opportunity with social need, then the total evaluation process must embody both aspects in an appropriate mix. Experts are generally best qualified to assess the opportunity for scientific progress, while broadly representative laymen in close consul-tation with experts may be best qualified to assess societal need. The optimal balance between opportunity and need can only be arrived at through a highly interactive, mutual education process involving both dimensions.

The original rationale of Vannevar Bush was to stress opportunity as the prime criterion of government support, assuming that the need-fulfillment aspect would take care of itself through the workings of the market or the political process after the results of the initial research were widely available. In other words, the function of government support would be primarily to push out the frontiers of opportunity and thereby generate a new technical capacity or resource on which society would be able to draw to fulfill present or future needs. Even though the oppor-tunities might overlap, the spectrum of needs would be so broad that most opportunities would prove relevant to some need—if not immedi-ately, then at some time in the future. The superior efficiency of the opportunity-oriented mode of approach would more than compensate for its less certain immediate relevance to an identified need. This is especially true because knowledge in the context of a well-developed conceptual framework can usually be communicated much more efficiently, and with lower transaction cost than knowledge in bits and pieces generated in the course of solving a technological problem.

The new paradigm, as articulated in a 1994 report by President Bill Clinton and Vice President Al Gore, is "Science, the Endless Resource" instead of "Science, the Endless Frontier."[44] At first sight, little difference is apparent, because in both cases a resource for solving societal problems is being generated. However, the new metaphor demands not only that the ability to create new knowledge be continually improved, but also that the ability to integrate new knowledge with old knowledge and enlist it in the betterment of the human condition be continually enhanced. This modification of the old Bush idea has been recently formulated by Paul A. David, David C. Mowery, and W. Edward Steinmueller in the following terms:

> The emphasis in science and technology policy has been placed on fostering the generation of new knowledge, rather than the distribution of knowledge and the possibilities of improving the performance of the system by improving access to the existing knowledge stock. We too are persuaded that this thrust has been maintained for too long, and there is a case to be made now for restoring some balance; in other words, to raise not only the marginal social rate of return on future R&D expenditures, but to increase the social payoff from such outlays made in the past by increasing the commercial exploitation of the knowledge.[45]

This is a genuinely new challenge to science policy and one toward which the new national policy is groping. The problem is that no understanding has emerged about what it means in practical terms for the conduct of the technical enterprise. David, Mowery, and Steinmueller suggest that

> the educational programs of those individuals that pursue a professional course of study in management need to incorporate an understanding of the nature of technology and the relation between technology and business. Similarly, those individuals pursuing

44. Clinton and Gore (1994, pp. 1–2).
45. David, Mowery, and Steinmueller (1994).

courses of study in the scientific and technical disciplines need an understanding of the legal and economic structure which will impinge directly on their careers. Lastly . . . American universities might be asked to . . . enhance the capacity of domestic business to monitor and benefit from timely information with regard to market developments, as well as technological changes, taking place in other countries.[46]

Unresolved Issues

A number of issues of future research policy and strategy have come to the fore with the damping down of the cold war, the emergence of a new set of public concerns, and the new political constraints engendered by the 1994 election.

Will the national laboratories or other free-standing nonprofit research institutions separate from universities be the principal locus of government support for addressing the competitiveness and economic performance missions, or will the main locus remain within or in close association with the research universities?

The national laboratories have recently been moving aggressively to embrace the new mission of technological innovation for economic performance. They have certain natural advantages in competing for these opportunities and powerful incentives to do so. They have large numbers of permanent professional employees covering a narrower range of disciplines than universities, because the teaching mission of universities requires them to have relatively smaller numbers of faculty in any one area of specialized expertise. Thus the large laboratories are in a better position to focus high-level, experienced talent on a single well-identified problem, and, especially in the large DOD and DOE laboratories, they are accustomed to such mobilization of talent by virtue of their past public missions. They are more comfortable with tight time deadlines and progress milestones, even though some individual scientists may have enjoyed considerable freedom within their general domain of expertise.

However, university people are more accustomed to pursuing even applied problems in much greater depth than might be strictly justified as cost-effective in the context of a tightly defined mission objective. Through this pursuit they may make a larger net societal contribution in

46. David, Mowery, and Steinmueller (1994).

the long run, and they tend to remain in contact with a much wider range of scientific developments and be more flexible in changing their agendas in response to newly appearing technical opportunities.

A fine balance thus is involved from society's standpoint in getting the most out of the advantages of each type of institutional culture. Careful monitoring is required to prevent immediate urgencies and competitive pressures from weighting project and program selection in favor of supporting the labs because they fall under the auspices of federal agencies in a sense that university groups do not and should not be expected to, given their wider social mission and their multiple constituencies and sources of funding.[47]

The new political climate in Washington, however, further complicates this choice. The new Republican majority is much more hostile to using government to help industry, especially large industry, and thereby intervening in the domestic competitive balance. In general the Republicans appear more friendly to basic research, while lacking a clear understanding of what it is. A good deal of hostility is evident toward universities as institutions and what is seen as their loose governance. Pressures to balance the federal budget are likely to result in measures, such as caps on indirect cost reimbursement, federal support for graduate students, the downsizing of government health care systems, and requirements for cost sharing, that will erode the institutional capacity of universities to respond to national needs. In the future, the maintenance and enhancement of this capacity is likely to depend on the development of new types of industry associations to support academic research of a largely public good nature that will replace the type of research supported in both universities and national laboratories by government during the cold war. This is a type of infrastructural research on which much industrial innovation has implicitly depended (and to a growing degree) in the past, but which most industry does not know it needs or will need until the research disappears.[48]

What are the implications for university priorities of the increased importance of the knowledge synthesis and diffusion functions as compared with the previous emphasis on the creation of new knowledge?

A consensus is developing among science policy analysts that the mission of research universities should give higher weight to knowledge

---

47. For a more detailed discussion of the future role of universities in the U.S. research system, see Brooks (1993, pp. 202–34).

48. As an example of this type of thinking, see Romer (1993).

synthesis, repackaging for use, and dissemination of new knowledge. This implies everything from radical changes in instruction methods based on findings in cognitive science to the development of hardware and software as aids to knowledge access and retrieval and the training of future scientists and engineers in its use. However, the practical implications of this new priority for the organization and procedures of universities and for the development of intermediary institutions between universities and knowledge-intensive practitioners delivering services to the public outside universities have not been worked out, or even much discussed. This is a serious gap in current science and technology policy debates.

What will be or should be the impact of new governmental priorities of national economic performance and competitiveness on the participation of foreign nationals in government-sponsored (unclassified) research at universities and other research institutions, both governmental and nongovernmental?

The productivity of the U.S. technical work force has become steadily more dependent on a continuously growing influx of scientists, engineers, and skilled practitioners originating from all parts of the world. Many come from abroad to complete their advanced education in the United States. Others are trained abroad and come to the United States because of the wide opportunities available for using their skills more creatively than is possible in their home countries, as well as better salaries and living conditions. This has been regarded as an almost unalloyed advantage for the United States in enhancing its capacity for technological innovation for economic growth and job creation. Nevertheless, many inconsistencies exist among the legal mandates of new government programs with respect to encouraging the attraction and participation of foreign nationals or foreign-owned corporations.

Furthermore, as the prosperity of the rest of the world has increased, a reverse brain drain has occurred of people who have received either advanced education or substantial hands-on technological experience— or both—in U.S. industry, making them an attractive human asset for acquisition by foreign competitors. The rise of technonationalism as a consequence of U.S. competitiveness problems threatens the open science culture of American universities. Debate and disagreement are heard as to net costs and benefits in the long run associated with the unique openness in human terms of the U.S. science and technology system, in the research universities, and, to some extent, throughout all the technical activities of American society.

The U.S. scientific community has a strong bias toward open science, but, because of rising criticism from some quarters—whether justified or not—the technical community as a whole may no longer be able to treat the subject with benign neglect. It may have to be dealt with as an important policy issue, worthy of the development of a well-thought-out rationale for an explicit national policy even if it turns out to be the present one. My own bias is that U.S. openness in science is a long-term asset that should be preserved and enhanced but is already being eroded from within. Openness is no longer an automatic assumption even within parts of academia and will need to be debated frankly. The effect of the new postelection political climate is unclear.

What kinds of ground rules, criteria, and procedures should be developed for the extent and form of participation of government agencies at both the state and federal levels in international cooperative research and educational projects?

The United States is acquiring a reputation as an unreliable and unpredictable partner in international collaborative projects in science and technology. So far the focus of policy attention has been on so-called megaprojects, projects requiring costly one-of-a-kind facilities or complex and expensive logistic support (as in the Antarctic program, or certain oceanographic or atmospheric joint ventures). However, megaprojects are also becoming an important issue in the case of collaborative international research institutions, not restricted to fundamental science. They can be an important issue for institutions to which the United States contributes only a small proportion of the funds required to operate a jointly agreed program but insists on its own exclusive unilateral peer review of the program as a condition of its financial participation. A consistent policy and ground rules need to be formulated for multinational peer review, with the agreement of all the participating governments to abide by the results under all but the most exceptional circumstances, to avoid loss of one of the main advantages of collaborative effort. In the new political climate, international cooperative activities in science that depend on government funds are likely to fare badly, with international science being the first to feel the budgetary ax. Such activities are likely to have to depend on private funding sources from both the nonprofit and corporate sectors, and here the outlook is somewhat more favorable. However, whether this will come close to replacing the role the U.S. government has played in the past seems doubtful.

Given that the primary goal of public sector investment in technology creation or diffusion is net job creation in the long run (that is, more new and better jobs than those displaced through technology), what are the implications for the selection of the most appropriate institutions and technologies for government support? By what processes will these goals be translated into selection criteria that command widespread support in the communities outside as well as inside the science and engineering communities?

In the mid-1990s, U.S. agencies lack a consistent, well-thought-out rationale for government support of projects that involve collaboration and cofunding with industry for the benefit of economic development and job creation. Sharing of costs between government and particular companies should imply a demonstrable public benefit that cannot be fully captured in the form of profits to the contributing companies. More rapid environmental improvement is one example of such a public benefit. But profit and new jobs are often interpreted as public benefits as well. Criteria need to be developed that can lead to reasonably objective discrimination between alternative public investments in different projects that include private companies as partners. The development of such a rationale seems far away, with the consequent hazard of steadily increasing politicization of the choices as programs grow. In the post-1994 political climate, the case for public benefit will have to be made much more cogently and concretely for private-public cooperative projects, if any of the present ones are to have any hope of survival.

How can criteria for public investment in new knowledge creation or diffusion be developed to provide assurance that public funds are not displacing potential private investment that is likely to be more cost-effective because of its greater responsiveness to markets? Put in another way, how can the net increase in effective national investment for a given public investment be estimated reliably?

Ideally, the benefit of public investment in technology has to be judged by the extent to which it can be demonstrated to increase the total net investment, making realistic allowances for displacement of private spending, including consideration of the relative efficiency in terms of knowledge diffusion and exploitation of public and private R&D.

What are the long-term implications of the priority now being given to the economic performance mission as a justification for public investment in science and technology and for the allocation of public resources among four classes of institutions: universities, government laboratories and FFRDCs, nongovernmental nonprofit organizations of various types, and private industry? To what extent, if any, does this new mission require that public funds be partially channeled through political subunits such as states, regional compacts among states, or even lower political levels?

During the entire period since the end of World War II the percentage allocation of R&D support among research universities, national laboratories (including both civil service laboratories and FFRDCs), and private industry has changed relatively little except for one sharp increase in university funding between the mid-1950s and mid-1960s. The question is whether this relative allocation is still appropriate to the new priority being accorded to the economic performance of the private sector. Recent federal initiatives and their projected expansion in the next five or six years suggest the possibility of some shift in favor of industry. However, this likely will chiefly compensate for the reduction of military development programs presently contracted out to industry. The new emphasis on Cooperative Research and Development Agreements (CRADAs) in major national laboratories also suggests the possibility of some shift of resources from universities to national labs, although this is by no means certain. Recent developments also indicate that states or regional state compacts may become more important players in determining the allocation of R&D resources associated with economic performance, analogous to the land grant system for agriculture, though not with the same degree of decentralization.

Does the scale and dispersion of the economic performance mission imply some degree of splitting up and possible dispersion of the large aggregations of technical capacity now represented by the major national laboratories?

This issue raises the question as to whether such large centralized organizations as national labs are still appropriate to the new economic performance mission, or whether the labs themselves may be split up into more or less independent self-governing units, while perhaps retaining their geographical proximity to take advantage of scale economies in support services. This seems more likely than actual geographical dispersion of national lab-type activity because modern communications facilitate close relations with clients and collaborators, while geographical shifts entail serious difficulties in the context of the regionally based U.S. political system. Parts of the national laboratory system likely may be candidates for privatization, for which the present CRADA arrangements may be a sort of halfway house.[49]

The breaking up and closure of government laboratories is much more imminent in the new political climate than before the 1994 election. The

49. For an explanation of Cooperative Research and Development Agreements (CRADAs) and other matters relating to national labs, see Branscomb (1993, pp. 103–34).

luxury of waiting to see how things work out and to optimize gradually may no longer be an option. The great technical capabilities that these laboratories represent are likely to be lost in the absence of a viable and well-rationalized plan for their restructuring in some new configuration.

What problems does the emerging prospect of a rapid downsizing of the U.S. government-funded R&D system raise for the institutions most directly involved with the training of scientists and engineers, and what near-term changes may be required in the nature of their training and career expectations?

Even if the funding for university research were not to shrink rapidly—an unlikely contingency—the effect on the educational mission of the universities would be far reaching because of the sudden lack of scientific and technical jobs outside of academia and the nature of the external job market into which the majority of advanced students go. However, the present situation is so uncertain that speculation is difficult. The United States could conceivably be facing a situation with respect to its highly trained personnel that would look for a few years not totally dissimilar to that now faced by scientists and engineers in the former Soviet Union. The private sector is unlikely to absorb these people except very slowly because most of what industry does requires a large downstream investment relative to R&D in contrast to activities in defense, space, health, and environmental management—the government functions toward which most present government-funded R&D is directed. The private sector probably would not soon be in a position to support this downstream investment on a self-sustaining basis even if it greatly expanded its R&D investments. A tremendous and rapid step-up would be required in the R&D-intensity of the U.S. economy analogous to what happened in the World War II economy, which, however, was financed by government. A look at the world's problems that need to be attacked could theoretically justify such a step-up, but it seems highly unlikely politically except over a long period of structural adjustment. Such a step-up did occur in the 1970s, in the period of the energy crisis, but in retrospect the R&D investment of that period appears to have been inefficient and wasteful.

Comment by E. J. Moniz

I am pleased to bring the perspective of a practicing scientist to this discussion of the social contract. My perspective is shaped by my experience as a physicist at the Massachusetts Institute of Technology, a university with a strong focus on science and technology and an especially strong tradition of industrial collaboration. In general I believe it is important for science policy analysts and science policymakers to maintain strong links to and a firm foundation in the research community.

What is the working university scientist's view of core responsibilities in the social contract? First and foremost, frontier research must be done, research that adds substantively to the knowledge resource base and occasionally breaks through to a significantly different understanding of the world. Teaching is another responsibility, principally using the apprenticeship mode for advanced training. University scientists should be engaged in service—to the government, to industry, to the public. Those core responsibilities carry on through periodic reexaminations of the contract.

The contract also is interpreted as requiring societal support for certain conditions that are critical to the pursuit of frontier basic research. One is a commitment to excellence, as judged through peer review. This commitment is not sustained automatically in a democratic society. Another condition, particularly important in a university setting, is freedom of inquiry, which entails pursuit of issues driven by the intellectual imperatives of a scientific discipline. Such research is clearly only part, but a very important part, of the overall research and development (R&D) portfolio. Both conditions are under considerable pressure. In addition, the basic disciplinary structure of academic science may be under pressure. This organization has traditionally offered an important framework for excellence while, contrary to some opinions, still providing opportunities for multidisciplinary work. I am unaware of alternate schemes of university organization that offer the same degree of quality control.

The growing tension between agenda-driven (strategic) and discipline-driven (curiosity-driven) research has a long history and is healthy if not elevated to a focal point of the national science policy debate. All the federally supported R&D is presumably viewed as supporting national goals, broadly defined. As set forth in the Clinton-Gore statement *Science in the National Interest*, these goals include leadership in basic science and education through forefront research, as well as the knowledge base for future technologies. However, even with a narrower definition of national goals, only about 10 percent of the federal basic research investment (that is, federal R&D, excluding applied research and development) is aimed at scientific knowledge with little bearing on federal missions (even there, the direct technology developments, such as accelerator technology in subatomic physics research, can be of very immediate consequence). And that research is generally an important part of the tightly interwoven scientific fabric.

The focus on the National Science Foundation (NSF) in the strategic research debate distorts this picture. The NSF-supported research programs are a critical part of the overall portfolio, particularly for university-based research. However, about three-quarters of the federal support for the physical sciences resides with mission agencies, particularly the Department of Defense, the Department of Energy, and the National Aeronautics and Space Administration (NASA). And the vast majority of support in the health-related and life science areas is explicitly mission-driven. Thus the entire cross-agency portfolio must be considered in reexamination of the social contract. That portfolio is already overwhelmingly strategic, in the sense that basic research support is seen from the agency perspective as relevant to a long-term mission just as the investigator may be motivated primarily by disciplinary considerations.

The mission agency support of basic research is important for reasons other than the obvious one of increasing opportunity. For the agency, the basic research support and strong university connection, with its foundation in demanding peer review, can help sustain quality control across many activities. For the university researchers, the advantage often exists of institution-building, which is less readily available in the individual investigator, who usually receives NSF support. That is, when supported by a mission agency, researchers are basically doing the agency's work; at the NSF researchers compete for funds to do their own work. I do not wish to push this point too far in terms of practice. Nevertheless, this distinction is reflected in the level of ongoing commitment to an institution.

This diversity of approaches, to be distinguished from simply a diversity of sources, is a significant strength of the system. Even within the NSF, a variety of funding mechanisms are available. For example, at MIT, the NSF-supported Center for Material Science and Engineering is one of the most successful multidisciplinary activities. The NSF supplies an umbrella grant through peer-reviewed competition under the ground rules that faculty from different departments join together for specific purposes. Many, such as one in funny fluids (liquid crystals, gels, and so on.), have produced first-class basic research (phase transition phenomena), several startup companies, and interesting applications (for example, work on eye cataracts as a phase transition). This diversity of research styles and the broad support of agency missions are the attraction of the multiple-agency portfolio of basic research support. It enhances the connection of basic research with the national interest.

Although the continuing debate about strategic research would benefit from a broader perspective, other policy areas need more attention—for example, capital budgeting. The lack of originality associated with this suggestion does not necessarily diminish its importance or timeliness. The nonexpert is struck by the idea of capping discretionary spending without imposing discipline on the much larger entitlement budget. The labels "discretionary" and "entitlement" are themselves not helpful; although perhaps equally imprecise, labels such as "investment" and "operating" come closer to capturing the truth. Arbitrarily limiting one's investments for the future but not controlling operating costs seems backwards. I recognize the difficulty of defining investments and of sustaining discipline, but an intellectual response to the issue must be distinguished from a political one.

Specific development would surely be affected by more extensive use of capital budgeting. Considerable debate presumably would take place over what part of the R&D budget fits in the investment category. One area that is likely to become problematic and is clearly an investment is the development of large scientific facilities essential for frontier research in many disciplines. These long-term ventures could easily span a quarter century and would encompass R&D, design, construction, and operation. Front-loading much of the cost as an operating expense is prejudicial. The budget environment will make providing these essential capabilities increasingly difficult. The U.S. science leadership position will not be sustained without a national long-term planning and budgetary process that advances the development of major facilities along with other research and education support and taming of the budget deficit.

Another part of the social contract is concerned with the responsibilities of scientists in graduate training. Plenty of room exists for improvement outside of the primary task of guiding doctoral research. A certain priest-class mentality, while far from uniformly prevalent (and not uniformly deserved) and indeed diminishing, undoubtedly still serves many students poorly. The career outlook of many students is shaped in a way that narrows their future contributions to science and to society. This mentality can also lead to magnified attention to Ph.D. employment, when the issue is more one of expectations. For example, in the research area of nuclear physics, the employment pattern remains stable: basically full and useful employment, with comparable representation in the academic, national laboratory, and industrial (including medical applications) sectors. I am constantly surprised to find former students enthusiastically and very effectively pursuing problems far from their doctoral research. On just one day during a recent visit to Los Alamos, I came across a former nuclear experimental physics student who had just completed measurements over the Pacific Ocean on ocean-atmosphere couplings and a former quantum field theorist advancing quantum cryptography. These activities need to be celebrated as important and satisfying directions for students, and some concrete reforms need to be implemented, while preserving the depth of a frontier research experience as the defining element of doctoral work. For example, at the MIT physics department, the options for satisfying the Ph.D. breadth requirement are being enlarged, from a narrow set of courses in neighboring physics subdisciplines to a varied menu of other science and engineering courses, industrial internships, and multidisciplinary courses specifically developed in partnership with other departments. Progress is apparent, and its pace is affected by the broad policy discussions going on nationally. Much remains to be done; for example, teachers must be enfranchised as part of the research community.

The mode of support of graduate students also needs reexamination in the context of a broader human resources policy for science. A heavy reliance on research assistantships ties student support to the best research. In contrast, fellowships are tied to the best students, and traineeships, roughly speaking, are tied to the best programs. Traineeships can be tied much more closely to larger objectives in the research agenda and can be used more easily to evaluate institutional performance (in both research and education). The physical sciences could do with somewhat increased emphasis on traineeships.

References

Alic, John A., and others. 1992. *Beyond Spinoff: Military and Commercial Technologies in a Changing World.* Harvard Business School Press.

Barber, Bernard. 1962. *Science and the Social Order.* New York: Collier Books.

Bernal, J. D. 1958. *World without War.* London: Routledge & Kegan Paul.

———. 1978. *The Social Function of Science.* London: Routledge & Kegan Paul.

Boretsky, M. 1973. *U.S. Technology Trends and Policy Issues.* Program of Policy Studies on Science and Technology, George Washington University.

Branscomb, Lewis M. 1993. *Empowering Technology: Implementing a U.S. Strategy.* MIT Press.

Brooks, Harvey. 1968. *The Government of Science*. MIT Press.

———. 1973. "What Happened to the Technology Gap?" *Professional Nutritionist* (Winter): 1–11.

———. 1974. "Are Scientists Obsolete?" *Science* 186 (November 8): 501–08.

———. 1977. "The Dynamics of Funding, Enrollments, Curriculum, and Employment." In *Proceedings of American Physical Society Conference on Changing Career Opportunities for Physicists*. New York: American Institute of Physics.

———. 1980. "Basic and Applied Research." In *Categories of Scientific Research*. NSF 80-28. National Science Foundation.

———. 1985. "Technology as a Factor in U.S. Competitiveness." In *U.S. Competitiveness in the World Economy*, edited by Bruce R. Scott and George C. Lodge, 328–56. Harvard Business School Press.

———. 1986a. "The Research University: Centerpiece of Science Policy?" Working Paper WP 86–120. Ohio State University, College of Business (December).

———. 1986b. "National Science Policy and Technological Innovation." In *The Positive Sum Strategy: Harnessing Technology for Economic Growth*, edited by Ralph Landau and Nathan Rosenberg, 119–67. Washington: National Academy Press.

———. 1990. "Lessons of History: Successive Challenges to Science Policy." In *The Research System in Transition*, NATO ASI Series, Series D: Behavioral and Social Sciences, vol. 57, edited by Susan E. Cozzens, Peter Healey, Arie Rip, and John Ziman. Dordrecht: Kluwer Academic Publishers (Proceedings of Managing Science in a Steady State: The Research System in Transition. NATO Advanced Study Institute, Il Ciocco, Italy, October 1-13, 1989).

———. 1993. "Research Universities and the Social Contract for Science." In *Empowering Technology: Implementing a U.S. Strategy*, edited by Lewis M. Branscomb, 202–34. MIT Press.

Brooks, Harvey, and Eugene B. Skolnikoff. 1975. "Science Advice in the White House? Continuation of a Debate." *Science* 187 (January 10): 35–41.

Bush, Vannevar, and others. 1960. *Science: The Endless Frontier*. Reprint. National Science Foundation.

Clinton, William J., and Albert Gore, Jr. 1994. *Science in the National Interest*. Executive Office of the President, Office of Science and Technology Policy.

Cohen,Wesley, Richard Florida, and W. Richard Goe. 1994. *University-Industry Research Centers in the United States*. Carnegie Mellon University, H. J. Heinz School of Public Policy Management, Center for Economic Development.

Dasgupta, Parta, and Paul A. David. 1992. *Towards a New Economics of Science*. CEPR Publication 320. Stanford University, Center for Economic Policy Research.

David, Paul A., David C. Mowery, and W. Edward Steinmueller. 1994. "University-Industry Research Collaborations: Managing Missions in Conflict." Paper prepared for CEPR/AAAS Conference on "University Goals, Institutional

Mechanisms, and the 'Industrial Transferability' of Research." Stanford University, Center for Economic Policy Research, and the American Academy of Arts and Sciences, March 16–20.

Dertouzos, Michael L., Richard K. Lester, and Robert M. Solow. 1992. *Made in America: Regaining the Productive Edge*. Cambridge: Harper Perennial.

Doctors, Samuel I. 1971. *The NASA Technology Transfer Program: An Evaluation of the Dissemination System*. Praeger Publishers.

Dupree, A. Hunter. 1957. *Science in the Federal Government*. Harvard University Press.

Florida, Richard J., and Martin A. Kenney. 1990. *The Breakthrough Illusion: Corporate America's Failure to Move from Innovation to Mass Production*. Basic Books.

Freeman, Christopher. 1993. *The Economics of Hope: Essays on Technical Change, Economic Growth, and the Environment*. London and New York: Pinter Publishers.

Galbraith, John Kenneth. 1994. *A Journey through Economic Time: A Firsthand View*. Houghton Mifflin.

Hollomon, J. Herbert, and Alan E. Harger. 1971. "America's Technological Dilemma." *Technology Review* 31 (July/August).

Illinois Institute of Technology. 1968. *Technology in Retrospect and Critical Events in Science (TRACES)*. National Science Foundation.

Isenson, R. S. 1969. "Project Hindsight: An Empirical Study of the Sources of Ideas Used in Operational Weapons Systems." In *Factors in the Transfer of Technology*, edited by W. H. Gruber and D. G. Marquis, 155–76. MIT Press.

Lee, Denis M. S. 1992. "Management of Concurrent Engineering: Organizational Concepts and a Framework of Analysis." *Engineering Management Journal* 4 (June): 15–25.

McGraw-Hill Yearbook of Science and Technology. 1963. McGraw-Hill.

Mowery, David C., and Nathan Rosenberg. 1982. "The Influence of Market Demand upon Innovation: A Critical Review of Some Recent Empirical Studies." In *Inside the Black Box: Technology and Economics*, edited by Nathan Rosenberg, 193–241. Cambridge University Press.

National Academy of Sciences. 1974. *Science and Technology in Presidential Policymaking: A Proposal*.

National Science Foundation, Directorate of Mathematical and Physical Sciences. 1994. *Connections: Investments for the Twenty-First Century*. NSF 94–74.

National Science Foundation, National Research Council, Committee on the Social Sciences. 1978. *Social and Behavioral Science Programs in the National Science Foundation: Final Report*. National Academy of Sciences.

Romer, Paul M. 1993. "Implementing a National Technology Strategy with Self-Organizing Industry Investment Boards." *Brookings Papers: Microeconomics* (2): 345–99.

Sandelin, Jon. 1994. "University Patenting and Technology Licensing as a Mechanism of Technology Transfer." Paper prepared for CEPR/AAAS Conference on "University Goals, Institutional Mechanisms, and the 'Industrial Transferability' of Research." Stanford University, Center for Economic Policy Research, and the American Academy of Arts and Sciences, March 16–20.

Sapolsky, Harvey M. 1990. *Science and the Navy: History of the Office of Naval Research.* Princeton University Press.

Schilling, W. R. 1964. "Scientists, Foreign Affairs, and Politics." In *Scientists and National Policy Making,* edited by R. Gilpin and C. Wright. Columbia University Press.

Sherwin, Chalmers W., and others. 1966. *First Interim Report on Project Hindsight (Summary).* Department of Defense, Office of the Director of Defense Research and Engineering.

Smith, Bruce L. R. 1990. *American Science Policy since World War II.* Brookings.
———. 1992. *The Advisers: Scientists in the Policy Process.* Brookings.

Social Science Research Council. 1969. *The Behavioral and Social Sciences: Outlook and Needs.* Prentice-Hall.

U.S. Department of Commerce, Office of the Assistant Secretary for Science and Technology. 1979. *Domestic Policy Review of Industrial Innovation.* PB–290403–417.

U.S. House of Representatives. Committee on Science and Technology. 1980. *Analyses of President Carter's Initiatives in Industrial Innovation and Economic Revitalization.* 96 Cong. 2 sess. Government Printing Office.

U.S. Senate. 1970. Hearings before the Subcommittee on Science and Astronautics. 91 Cong. 2 sess. Government Printing Office.

Vernon, Raymond. 1966. "International Investment and International Trade in the Product Cycle." *Quarterly Journal of Economics* 80 (May).

CHAPTER 3

Science, Economic Growth, and Public Policy

Richard R. Nelson and Paul M. Romer

MAJOR LONG-TERM policy changes often flow from decisions made in times of stress. Thus choices about how to fight World War II ultimately led to the large-scale sustained support of university research by the federal government that has lasted over the half-century since World War II. The economic threat faced by the United States today is less acute than the security threat faced then. Nevertheless, it may lead to a fundamental realignment of science and technology policy and a major change in the economic roles of the university.

In the midst of the debate about how government support for science should be structured after the war, Vannevar Bush prepared his famous report *Science: The Endless Frontier*. Although the specific institutional recommendations from the report were not adopted, it set the terms for the subsequent intellectual debate about science policy. In a forthcoming analysis of Bush's report, Donald E. Stokes notes that Bush advocated government support for the kind of abstract science done by someone such as Niels Bohr, the physicist who played a pivotal role in the development of quantum mechanics.[1] Bush argued that public support for that kind of science would lead to advances in the work done by someone such as Thomas Edison, who takes existing knowledge and puts it to commercial use. (The argument comes from Bush, but the examples of Bohr and Edison come from Stokes.)

A decisive shift has occurred in the government attitude toward direct instead of indirect support for the Edisons. If Edison were alive today, setting up General Electric, he could apply for direct grants from government programs such as Small Business Innovative Research (SBIR), the

1. Stokes (forthcoming).

Advanced Technology Program (ATP), and the Technology Reinvestment Program (TRP). He could pursue Cooperative Research and Development Agreements (CRADAs) with the national laboratories. He could form a consortium of for-profit firms and get government matching money to develop a specific technology such as flat panel display screens. The government would also be much more willing to help him establish commercially valuable intellectual property rights over any fundamental discoveries that he might make. Some policymakers would encourage him to patent the sequence data on gene fragments, the scientific and practical importance of which no one yet understood.

As many students of science and technology have pointed out, good reasons exist to be dissatisfied with the linear model of the relationship between science and practical technology that is implicit in Bush's report. According to this now discredited model, the government merely puts resources into the Bohr-end of a production line and valuable products come out at the Edison-end.[2] However, equally good reasons can be cited to be worried about a strategy that sharply shifts government policy toward direct support of research and development (R&D) in industry, giving government money to Edison-like activities and strengthening property rights across the board. And the reasons for concern are amplified if such a policy shift means public support for basic research at universities dries up.

One important limitation of the linear model, and the one to be focused upon here, is that it is blind to basic research undertaken with practical problems in mind—work in which the Bohrs are directly motivated to lay the scientific basis for the work of the Edisons. In the map laid out by Stokes, such work is epitomized by the research of Louis Pasteur, a scientist whose research was primarily guided by practical problems but led him to explore fundamental scientific questions. Basic economic analysis suggests that different institutional arrangements be used to support the work of a Bohr and an Edison, but the example of Pasteur indicates that strong linkages between the two are desirable. Both of these kinds of work are more productive when they rub up against each other.[3]

American universities have been uniquely successful in promoting this kind of interaction. Before World War II, they did this by catering to the needs of the private business sector. They provided the home for new

2. See Kline and Rosenberg (1986).
3. Rosenberg (1982).

scientific fields such as metallurgy, which was developed to advance steel-making technology. After the war, U.S. universities became world-class centers of Bohr-style science, but they also gained new strength in the Pasteur-like activities. In large part, this occurred because government agencies such as the National Institutes of Health and the Department of Defense provided massive support for what came to be called mission-oriented basic research.

The set of practical problems that animates Pasteur-style science within the university could now be adjusted. Emphasis on problems in the areas of defense and health could be reduced, and attention to the broad range of scientific and technical challenges that arise in the private sector could be increased. This change could be implemented without endangering the national strength in Bohr-style science, without trying to privatize Pasteur-style science, and without creating strong property rights that could impede the free flow of knowledge that is generated by this work. The preservation, with reorientation, of Pasteur-style science within the university will both strengthen Bohr-style science and help meet the changing practical demands being put on science.

This may not be adequately understood. Instead of offering new and different opportunities for the Pasteurs of the university, policymakers may try to convert both the Bohrs and the Pasteurs into Edisons. Fearful of this prospect, the Bohrs and Pasteurs may fight any proposal for readjustment. Government leaders may therefore bypass the university in frustration and fund the Edisons of the private sector directly. Over time, the work that was done by Pasteurs in the university will be shifted to the private sector through a combination of direct grants, matching money, and stronger property rights, where it will become Edison work. The Bohrs may acquiesce in this privatization and eventual destruction of Pasteur-style science because it buys them protection from political demands for changes in their part of university research. The result could be the kind of separation that has been avoided until now, with Bohrs working in isolation from the Edisons, and with little work in Pasteur's quadrant. Lost would be the unique features that made U.S. universities so successful in generating good science and strong economic growth.

The Current Policy Context

Before describing what economists know about the connections between science, technology, and economic growth, the economic context of the current debate about science policy should be explained. This context has been shaped by the erosion of the large and widespread technological and economic lead that the United States had over other countries during the 1960s and the worldwide slowdown in income and productivity growth since the early 1970s. The erosion of the U.S. lead is easy to explain and probably was inevitable, but the slowdown in growth is not well understood.

The post–World War II economic and technological dominance of the United States was the consequence of two distinct waves of economic growth.[4] The first wave, which dates from the late nineteenth century, began when U.S. universities were not strong centers of scientific research. The act that created the land grant college system in the 1870s described the mission of these institutions as the development of the agricultural and mechanical arts. The research that did take place tended to reflect this strong practical orientation. European intellectuals were disdainful of the vocational orientation of American universities. And as late as the 1930s young American scientists who wanted advanced scientific training generally went to Europe to get it.[5]

The early U.S. successes in such industries as automobiles and steel were not the result of any particular American strength in science. Instead, firms here achieved dominance in the techniques of mass production in large part because they operated in the world's largest common market. They had access to many consumers and to ample supplies of inexpensive raw materials. But universities also played an important role. Because of the unusual practical orientation of the U.S. system of higher education, U.S. industry had access to a large pool of well-trained engineers and was able to develop professional managers to a far greater degree than was the case in Europe.

The second major wave of American economic success was in high-technology industries. These developed after World War II and were made possible by rapidly developing American capabilities in science. World War II was a watershed in American science and technology in several respects. After the war, the federal government became the principal patron of university research. By the middle 1960s the American university research system had become the world's best across a spectrum that included almost all fields of science. This improvement in the quality of American science was accompanied by major procurement and indus-

4. Abramovitz (1986); Nelson and Wright (1992).
5. Rosenberg and Nelson (1994).

trial R&D programs of the Department of Defense and, for a period of time, the National Aeronautics and Space Administration (NASA). These programs created the initial market for some of the high-technology goods that made the first use of the rapidly developing body of scientific knowledge. However, in many cases, the market for high-technology goods drew forth the science that made these goods possible.

Increased government support for science was accompanied by two other developments. First, a large number of young Americans earned a university degree. While only a small fraction of college majors were in the natural sciences or engineering, the sheer numbers of Americans receiving undergraduate and postgraduate training meant that by the late 1960s the fraction of scientists and engineers in the U.S. work force stood well above that in Europe and Japan. Second, both private and public monies flowing into industrial R&D increased greatly. By the late 1960s the U.S. ratio of industrial R&D to gross national product (GNP) was far higher than in any other country.[6] All these factors combined to give firms in the United States a commanding position in such high-technology fields as computers, semiconductors, aircraft, and pharmaceuticals.

By the late 1960s American economic dominance was beginning to erode, as Japan and the advanced industrial nations of Western Europe began to catch up. Two basic factors were behind this process. First, the economies of the industrialized nations integrated rapidly. Reductions in transport costs and the removal of trade restrictions meant that manufactured products and raw materials could move more readily between countries. In addition, increased flows of direct foreign investment let firms from the United States put their knowledge and technology to work in many other countries. Second, other countries made investments in science and engineering education and in research and development. Together these developments made achieving rough parity with—and in some cases, going beyond—the United States in traditional areas of mass production possible for several countries. The U.S. high-technology industries, however, have generally continued to do well in the face of strengthening foreign competition.

Most economists believe that convergence among the advanced industrial nations was inevitable. In a world where transportation and communication costs are falling and where governments are removing artificial barriers, the same forces that operate within the borders of the United States will operate between countries. At the time of the Civil

6. Nelson and Wright (1992).

War, economic activity in the South was very different from that in the industrialized Northeast. Because of the greatly increased mobility of goods and firms that has been the result of advances in transport and communication technology since then, economic activity in the two regions now looks much the same.[7]

At the same time that convergence between the industrialized nations occurred, productivity and income growth slowed significantly from the pace it had achieved during the quarter century after World War II. This slowdown was evident first in the United States but was also apparent in the other industrialized economies. Economists are still uncertain as to what lay behind the global slowdown beginning in 1970 or, to put the matter in another way, why growth that proceeded at unprecedented rates during the 1950s and 1960s has returned to levels that are closer to historical norms.[8] In any case, economists are nearly unanimous in holding that the rapid growth of other nations was not a cause of the slowdown in growth in the United States.

Nevertheless, the combination of convergence and slow growth created a public perception that the United States is suffering from a serious relative decline in its economic performance. As a result, the nature of the policy discussion in the United States changed regarding the appropriate role of the government in supporting technology and science. The loss of the dominant position held by American firms has caused the policy discussion to focus on measures that could enhance their competitive position. The productivity slowdown, which manifests itself most dramatically in stagnation of the wages paid to low-skilled workers, has generated additional support for government measures that would directly spur economic growth.

The slowdown also has meant that government revenues have not grown as rapidly in recent years as they did during the 1950s and 1960s. The slowdown in the rate of growth of private income has increased political resistance to higher tax rates. As a result, political support for the strategy of dealing with national problems by spending public money has fallen. Also, as seems always to be the case when times get harder, disenchantment has emerged with government policies and programs that were widely regarded as appropriate and efficacious during earlier, better economic times.

One important manifestation has been growing dissension about

7. Barro and Sala i Martin (1991).
8. Baily and Chakrabarti (1988); Griliches (1994).

whether the large-scale U.S. government support for basic research, primarily at universities, is worth the cost. Increasingly, suggestions are made that university research support ought to be more closely targeted on areas and activities that are deemed likely to feed directly into technological innovation. This dissatisfaction has influenced the design of the new technology programs. Except in the area of defense procurement, the government traditionally has used the university as an intermediary when it wanted to encourage economic and technological development in the private business sector. Many new technology programs largely bypass the university, directly influencing research activity within firms and, for the first time, in areas where the federal government will not be the primary user of the goods being developed.

Several other factors further complicate the situation for universities. The end of the cold war poses a serious threat to existing defense-related support for university research in fields such as electrical engineering, computer science, and materials science. Growing concern about health care costs may soon threaten research support for the biomedical sciences. An increasing number of young scientists who had expected to follow an academic career are finding that path blocked by a lack of jobs. Universities are responding to the feared cutbacks in government research funding by soliciting more support from industry.

At the same time that public support of university basic research has come under attack, some of the private organizations that did pathbreaking basic research—Bell Labs, IBM, Yorktown, Xerox PARC—have been cutting back on expenditures or reallocating their energies to projects that have quicker payoffs or where the results more easily can be kept proprietary. Some of these same companies also are pulling away from their previous support of academic research.

The current debate about government support for science and technology reflects all this. Decisions made now will determine how scientific research in universities and technological development in industry will evolve, perhaps for decades to come. Behind every position in this debate lies a set of assumptions about the relationships among science, technological innovation, and economic growth.

Technology and Economic Growth

From the beginning, economists have appreciated the importance of technical advance. One of the most striking parts of Adam Smith's pioneering analysis of economic principles, *The Wealth of Nations*, was his famous description of productivity improvement in the making of pins. A good part of that description involved technical advances.

Technological advance was seen as the force that could offset diminishing returns. Diminishing returns—the notion that the marginal benefits decrease as the effort in any activity increases—is fundamental to any explanation of how a market economy allocates resources. Classical economists reasoned as follows: The amount of food produced by each agricultural worker is very high when few workers are on a given area of land. Output per worker diminishes as more people work the given amount of land. The result is a pessimistic view of the prospect for sustained economic growth. As Thomas Malthus and others pointed out, in the absence of some offsetting influence, diminishing returns in agriculture implies that the output of food per person will fall as the population increases. The inevitable outcome would be famine and starvation.

By the end of the nineteenth century, this dismal forecast was evidently wrong. Population and food output had each increased dramatically. Economists observed that discovery and invention kept Malthus's prediction from coming true. With a fixed set of technological opportunities, the return in any activity diminishes. But over time, new techniques of production are introduced. Initially, these new activities offer high returns. As resources are shifted into them, the returns fall, but new discoveries and new techniques keep the process going.

Economists were preoccupied with other questions during the first half of the twentieth century, especially with macroeconomic stabilization because of the worldwide disruptions experienced during the interwar period. When economists returned to the study of long-run trends in the 1950s, both the empirical studies and the theoretical writings affirmed the importance of technical advance to economic growth. Technological change has a direct effect on growth by increasing the amount of output that can be produced with fixed quantities of capital and labor. Economists try to measure the direct effect with estimates of total factor productivity growth or the growth accounting residual. Early estimates attributed most of the growth in per capita income to this effect alone.[9] More recent estimates have attributed a larger fraction of growth to the accumulation of physical and human capital and a smaller fraction to technology.[10]

9. Abramovitz (1956); Solow (1957).
10. Jorgenson, Gollop, and Fraumeni (1987).

Technical advance also has an indirect effect because it raises the return on investments in physical and human capital. With no technological advance, diminishing returns would reduce the return on both of these types of capital to zero. Capital accumulation would stop. In a fundamental sense, all economic growth, even the growth directly caused by capital accumulation, can ultimately be attributed to technological change.

A second line of work tried to measure the rate of return on investments in technology. In one famous and revealing calculation, Zvi Griliches showed that the investments in agricultural research that produced benefits with hybrid corn were about seven times larger than the costs and yielded an internal rate of return of about 40 percent, which is much higher than returns on other forms of investment.[11] Other calculations found similar rates of return on research investments in other parts of agriculture and in manufacturing.[12] These estimates measure the social rate of return because the entity that does the research—either the government or the private firm—often fails to capture all benefits. In the jargon of the field, much of the benefit comes in the form of spillovers that are captured by others.

The existence of a differential between private and social returns is essential to understanding why high rates of return on research and development could persist. If all firms could capture all of the benefits and earn a 40 percent return on investment in R&D, many firms would increase their R&D investments. As they did, the return to research would be driven down to a more normal level. Because large returns on investment in research still are available, the inference is that private investors have difficulty capturing all of the benefits from their investments.

The divergence between the private and social return to R&D investment provides an important justification for policies that would encourage R&D. From the point of view of society, the income-maximizing strategy is to invest first in those activities that offer the highest rate of return. This criterion suggests that not enough is being invested in the activities that generate technological advance. To address the question of how this deficiency could be remedied, a precise understanding is needed of what these activities are and what the government can do to influence them.

11. Griliches (1958); Griliches (1992).
12. Mansfield and others (1977).

The Economics of Software

Although economists have long appreciated that technical advance is central to the process of economic growth, a complete understanding of the key processes, investments, and actors that combine to produce it has not come easily. These processes are complex and variegated. Economists broadly understand that the advance of technology is closely associated with advances in knowledge. Furthermore, new knowledge must be embodied in practices, techniques, and designs before it can affect economic activity. Beyond this, different economic analyses focus on or stress different things.

Some discussions stress the public good aspects of technology, seeing new technology as ultimately available to all users. Others treat technology as largely a private good, possessed by the company or person that creates it. Many economists have studied research and development as the key source of new technology. Those that have focused on R&D done by private, for-profit business firms assumed that the technology created through corporate R&D is, to some extent at least, a private good. In contrast, economists who have stressed the public good aspects of technology have focused on government investments in R&D, spillovers from private R&D, or both. (These spillovers are another manifestation of the divergence between public and private returns.) Still others argue that a single-minded emphasis on organized R&D as the source of technical advance is too narrow. They point to evidence that learning-by-doing and learning-by-using are important parts of the processes whereby new technologies are developed and refined.

Another matter on which economists have been of different minds is whether technical advance, and economic growth fueled by technical advance, can adequately be captured in the mathematical models of economic equilibrium developed to describe a static world. Joseph Schumpeter and economists proposing evolutionary theories of growth have stressed that disequilibrium is an essential aspect of the process. In contrast, recent theories that descend from neoclassical models presume that the essential aspects of technical advance and economic growth can be captured by extending the static equilibrium models.[13]

While the important open questions about how economists ought to understand technical advance should not be underplayed, a workable consensus for policy analysis seems to be emerging from these divergent

13. For a discussion of these models, see Romer (1994).

perspectives. Technology needs to be understood as a collection of many different kinds of goods, with the attributes of public goods and private goods in varying proportions. Some are financed primarily by public support for R&D, others by private R&D. Business firms and universities are involved in various aspects of the process. Other parts of technology are produced primarily through leaning-by-doing and learning-by-using, both of which can interact powerfully with research and development. Some aspects of the process are well treated by equilibrium theories, with their emphasis on foresight, stationarity, and restoring forces. Other aspects are better suited to the evolutionary models, with their emphasis on unpredictability and the limits of rational calculation.

One way to summarize this emerging view is to focus on three types of durable inputs in production. Imagery and language from the digital revolution allow us to describe three different types of inputs as hardware, wetware, and software. Hardware includes all the nonhuman objects used in production—both capital goods such as equipment and structures and natural resources such as land and raw materials. Wetware, the things that are stored in the wet computer of the human brain, covers the human capital that mainstream economists have studied and the tacit knowledge that evolutionary theorists, cognitive scientists, and philosophers have emphasized. Software represents knowledge or information that can be stored in a form that exists outside of the brain. Whether text on paper, data on a computer disk, images on film, drawings on a blueprint, music on tape—even thoughts expressed in human speech—software has the unique feature that it can be copied, communicated, and reused.

The roles of hardware, wetware, and software can be discerned in a wide variety of economic activities. Together they can produce new software, as when a writer uses his or her skills, word processing software, and a personal computer to write a book. They can produce new hardware, for example, when an engineer uses special software and hardware to produce the photographic mask that is used to lay down the lines in a semiconductor chip. When an aircraft simulator and training software are used to teach pilots new skills, they produce new wetware.

The three types of inputs can be discerned in activities that are far removed from digital computing. In the construction of the new city of Suzhou in mainland China, the government of Singapore says that its primary responsibility is to supply the software needed to run the city. The hardware is the physical infrastructure—roads, sewers, buildings, and so on—that will be designed according to the software. The wetware initially will be the minds of experts from Singapore but eventually will be supplied by Chinese officials who will be trained in Singapore to staff the legal, administrative, and regulatory bureaucracies. The software comprises all the routines and operating procedures that have been developed in Singapore, which range from the procedures for designing a road, to those for ensuring that police officers do not accept bribes, to instructions on how to run an efficient taxi service.

Traditional models of growth describe output as a function of physical capital, human capital, and the catchall category of technology. The alternative proposed here has the advantage of explicitly distinguishing wetware (that is, human capital) from software, an essential first step in a careful analysis of the intangibles used in economic activity. The next step is to identify the reasons that software differs from hardware and wetware.

Economists identify two key attributes that distinguish different types of economic goods: rivalry and excludability.[14] A good is rival if it can be used by only one user at a time. This awkward terminology stems from the observation that two people will be rivals for such a good; they cannot both use it at the same time. A piece of computer hardware is a rival good. So are the skills of an experienced computer user. However, the bit string that encodes the operating system software for the computer is a nonrival good. Everyone can use it at the same time because it can be copied indefinitely at essentially zero cost. Nonrivalry is what makes software unique.

Although a nonrival good can physically be used by many people, some may not be permitted to use it without the consent of the owner. The second property, excludability, thus comes in. A good is said to be excludable if the owner has the power to prevent others from using it. Hardware is excludable. To keep others from using a piece of hardware, the owner need only maintain physical possession of it. The American legal system supports each citizen's efforts to do this.

Making software excludable is more difficult because possession of a piece of software is not sufficient to keep others from using it. Someone may have surreptitiously copied it. The feasible alternatives for establishing some degree of control are to rely on intellectual property rights established by the legal system or to keep the software, or at least some crucial part of it, secret.

---

14. For textbook treatments of these concepts see Cornes and Sandler (1986), Starrett (1988), or Stiglitz (1988).

The legal system assigns intellectual property rights to some kinds of software but not others. For example, basic mathematical formulas cannot be patented or copyrighted. At the present time, scientists who develop algorithms for solving linear programming problems cannot claim intellectual property rights on the mathematical insight behind their creation. However, the code for a computer program, the text of a novel, or the tune and lyrics of a song are examples of software that is excludable, at least to some degree.

The two-way classification of goods according to excludability and rivalry creates four idealized types of goods. Private goods and public goods are two of them. Private goods are both excludable and rival. Public goods are both nonexcludable and nonrival. The mathematical principles used to solve linear programming problems are public goods. Because they are software, they are nonrival; copying the algorithms out of a book is physically possible. Because the law lets anyone copy and use them, they are nonexcludable.

The two other types of goods have no generally accepted labels but are important for policy analysis. The proverbial example of a good that is rival but not excludable is a common pasture. Only one person's livestock can eat the grass in any square foot of pasture, so pastureland is a rival good for purposes of grazing. If the legal and institutional arrangements in force give everyone unlimited access to the pasture, it is also a nonexcludable good. "The tragedy of the commons" illustrates one of the basic results of economic theory: Free choice in the presence of rival, nonexcludable goods leads to waste and inefficiency.

The opposite holds when nonexcludable goods are made excludable. The standard "invisible hand" theorem in economics says efficient outcomes result when people are free to trade and use private goods—the rival goods that are excludable. This contrast—between bad outcomes when rival goods are not excludable and good outcomes when they are— is the basis for a widespread presumption that strong property rights are better than weak property rights. The result for rival goods does not necessarily carry over to the category of nonrival goods such as software.

The final type is nonrival goods that are potentially excludable.[15] Society has a choice about excludability. It can establish and enforce strong property rights, in which case market incentives induce the production of such goods. Alternatively, it can deny such property rights.

---

15. See Romer (1990) for discussion of the importance of nonrivalry for understanding innovation and growth.

Then, if the goods are to be provided, government funding, private collaborative effort, or philanthropy is needed. Many of the most important issues of public policy regarding technical advances are associated with this choice. For rivalrous goods, establishing and enforcing strong property rights is generally a good policy (although exceptions exist). For nonrivalrous goods, the matter is much less clear.

By and large, society has chosen to give property rights to the kind of software commonly called technology and to deny property rights but provide public support for the development of the software commonly referred to as science. But the lines between technology and science are not always clear. And in any case, simply noting that a difference traditionally has been drawn in the treatment of property rights between these two types of software does not provide a reason for that difference. Copyright and patent protection could be denied for the software sold in computer stores or for new semiconductor designs, and other mechanisms, such as government subsidies for production, could be relied upon. In principle, patents could be granted to basic scientific discoveries, even to mathematical formulas. Reasonably effective mechanisms are available for ensuring that the owner of a song gets paid whenever it is broadcast on the radio or used in a movie. A similar system could be set up to establish property rights in many areas of science. And why not?

Establishing property rights on software enables the holder of those rights to restrict access to a nonrival good. When such restriction is applied—for example, by charging a license fee—some potential users for whom access would be valuable but not worth the fee will choose to forgo use, even though the real cost of their using it is zero. So putting a price on software imposes a social cost (positive value uses are locked out), and in general, the more valuable the software is to large numbers of users, the higher will be the cost. One example of how the choices of working scientists are influenced by costs is that some experiments using PCR (polymerase chain reaction) technology are not being done because the high price charged by the current patent holder makes this research prohibitively expensive. More experiments would be carried out if scientists could use the technology at the cost of the materials involved.

The situation is different from what is entailed by establishing property rights on rival goods. Only one entity can make use of a rival good, at least at any one time. Property rights, and an option to sell them, encourage the rival good to be used where it is most valuable.

The U.S. legal system takes account of the ambiguous character of property rights on software. Patents are issued for some discoveries, but they are limited in scope and expire after a specific period of time. For rival goods, this would be a terrible policy. Imagine the effects if the titles to all pieces of land lapsed after seventeen years. For some nonrival goods such as works of literature or music, copyright protection lasts much longer than patent protection. The rationalization is that costs from monopoly control of these goods creates relatively little economic inefficiency. For other goods such as scientific discoveries and mathematical formulas, the law provides no protection. This presumably reflects a judgment that the costs of monopoly power over these goods are too high, that society is better off relying on nonmarket mechanisms such as philanthropic giving and government support to finance and motivate the production of these types of software.

One important distinction between different types of software arises because of differences in the amount and variety of additional work that needs to be done before that software makes a contribution that appeals to final consumers.[16] Property rights on software that is directly employed by final consumers can lead to high prices—consider the high prices on some pharmaceuticals—and eliminate use by some parties who would value the software, but will not or cannot pay the price. However, for software that is close to final form, users can make reasonably well-based benefit-price calculations.

However, with software used primarily to facilitate the development of subsequent software, the uncertainties regarding value are compounded by the number of steps that need to be taken before something of final value can be achieved. As a result, the establishment of an efficient license market is difficult. Any market for software such as mathematical algorithms and scientific discoveries far removed from the final consumer would risk being grossly inefficient. Over time, many producers have to intervene, making improvements and refining the basic idea, before such software can be embodied in a technique, practice, or design that produces value and is sold to a final consumer. Economic theory says that the presence of monopoly power at many stages in a long and unpredictable chain of production can lead to severe inefficiency.

The problem is not just that the price of the software that is produced would be too high. The biggest loss may come from the types of software that are not produced but could have been. In the worst case, property rights that are too strong could preempt the development of entire areas of new software. In the computer software industry, for example, the

16. See, for example, Arrow (1962).

question to consider is, "What if someone had been able to patent the blinking cursor?" The point applies equally well to many other important discoveries in the history of the industry: the notion of a high-level language and a compiler, the iterative loop, the conditional branch point, or a spreadsheet-like display of columns and rows. Extremely strong property rights on these kinds of software could have significantly slowed innovation in computer software and kept many types of existing applications from being developed.

In the production of computer software, basic software concepts are not granted strong property rights. Software applications, the kind of software sold in shrink-wrapped boxes in computer stores, are protected. This suggests a simple dichotomy between concepts and final applications that mirrors the distinction between the search for basic concepts by a Niels Bohr and the search for practical applications by a Thomas Edison. As the work of Pasteur indicates, this dichotomy hides important ambiguities that arise in practice. At the extremes, the distinction between concepts and applications is clear, but in the middle ground no sharp dividing line exists. Courts are forced to decide either that software for overlapping windows or specific key sequences should be treated as essential parts of an application that are entitled to patent or copyright protection or that they are basic concepts that are not given legal protection. In the realm of software, there are many shades of grey. The simple dichotomy nevertheless serves as a useful framework for guiding the economic and policy analysis of science and technology, for science is concerned with basic concepts and technology is ultimately all about applications.

Science and Technology

One of the dangers of drawing sharp policy distinctions between basic concepts and applications arises because progress in the development of both types of software is most rapid when they interact closely. The ideal policy treatment of these two types is different, but if badly designed policies interfere with this interaction, they can do great harm.

Most important new technologies come into existence in an embryonic and imperfect form. In many cases, people have only a limited understanding of both the underlying basic concepts and of the range of possible applications. Some time and effort were needed after the discovery of the transistor at Bell Labs before transistors could be used in practical applications. Many years passed before the transistor evolved from its early freestanding state to when transistors were collected in integrated circuits, and many more for the development of higher density and faster circuits. Many researchers working in many different firms contributed to these developments. In the beginning, no one anticipated the many uses to which it would be put. If Bell Labs had had extremely strong property rights over the use of the transistor, many of the most important improvements in design and new uses for it might never have been discovered.

The story of the laser follows similar lines. AT&T, which had rights to the invention, at first could not see a way to use the laser in the communications business. Successive generations of the laser have turned out to have a wide range of applications, the vast majority of them outside the range of the telephone system. One important application, however, has been in fiber optics, which currently is revolutionizing that system.[17]

In the cases of both the transistor and the laser, technological development was marked by great uncertainty and considerable differences of opinion regarding how to make the technology better. Wide participation in the process of refinement and exploration was needed to produce the many applications that consumers now buy.

In most of the technologies whose development has been studied in detail, technical progress proceeded through a lengthy, complex evolutionary process. At any time, a number of actors attempted to develop variants or improvements on prevailing technology. They competed with each other and with prevailing practice. Some turned out to be winners; others, losers. The winners often enjoyed wide market success. At the same time, they provided a new base from which subsequent technological advance, often made by others, could progress.

Most innovations that arise in the private sector are a mixture of new concepts and applications that are ready for sale. Successful inventors can make a profit, at least for a time, on the sale of applications, because they generally are protected. Even if the legal system does not provide effective protection, the advantages of acting first and secrecy are often enough to let someone earn a profit by selling a new application. In almost all cases, the basic concepts became public software, available for

17. Rosenberg (1994).

the rest of the technological community, both in the private sector and in the university to build on.[18]

Strong property rights that interfered with widespread participation would reduce the diversity in the evolutionary process and slow progress. But weak property rights create spillovers. They reduce the private incentives for doing research and induce a divergence between the social and private rates of return to research. An effective social system for inducing technological progress will therefore tolerate weak property rights on basic concepts but will subsidize some types of research to offset the tendency for research effort to be too low. Because the search for concepts and the search for applications can lead both to important new discoveries, both are candidates for subsidies. Since World War II, a significant portion of the subsidies in the United States took the form of unrestricted support for university research into basic concepts (as provided, for example, by the National Science Foundation), but an even larger fraction was devoted to support for research in basic concepts that were relevant for practical applications in the areas of defense and health.

Before the war, the government provided research support in the field of agriculture and private philanthropic organizations funded some areas of basic science research. However, the bulk of the subsidies were directed at training scientists and engineers, most of whom went to work in the private sector. Some of this support came from the federal government, through its land grants to states. Some came from the operating budgets of the states themselves. Important support also came from the philanthropic activity of people such as George Eastman and Arthur D. Little (who helped create chemical engineering at the Massachusetts Institute of Technology) or organizations such as the Carnegie Foundation and the Rockefeller Foundation (which fostered the development of physics, the social sciences, and molecular biology).[19]

For the laser and the transistor, fields of scientific study grew up around the new technologies. The advent of the transistor provided a whole new agenda for research for electrical engineering and material science. The laser has had a major effect in fields such as physical chemistry and revitalized the field of optics. These scientific fields worked backward from applications and tried to uncover the basic concepts that helped explain how and why they worked.

In both cases, the original inventions drew extensively on scientific

18. Levin and others (1897).
19. Kevles (1979); Kay (1993).

knowledge. After their achievement, the technologies themselves became the subject matter of scientific research. In turn, the growing body of scientific understanding about the technologies provided important inputs into their refinement and further development.

Technological progress was rapid before and after World War II, in environments that provided different kinds of support for science and technology. The history of specific technological areas shows that the development of basic concepts and applications are intimately intertwined. Asking whether applications or basic concepts is the prime mover in generating scientific and technological progress is pointless. Because each can encourage the other, neither can be singled out. This has not, however, stopped people from trying.

In the 1950s and 1960s, scholars studying technical advance debated the relative importance of perceptions of demand or opportunities opened by science. Implicit in the debate were two views about policy options for stimulating technical advance and economic growth. The interpretation based on scientific opportunity was associated with a science-push policy: Support scientific research, and the economic and technical benefits will follow. The perceptions of the demand view suggested that measures designed to increase economic activity in the private sector should be the highest priority.

A number of studies concluded that the key factor that motivated the initiation of particular projects was almost invariably a perception of a demand. Studies such as that by M. Gibbons and R. Johnston have documented that scientific understanding and techniques often played a critical role in successful inventive efforts, but that the understandings and techniques drawn upon often were relatively old.[20] A study funded by the Department of Defense, Project Hindsight, explored the key scientific and technical breakthroughs that allowed the development of a number of important weapons for the military, and it found that almost invariably these came about as the result of research that was addressed to particular needs, not basic research done with little awareness of or concern about those problems.

The National Science Foundation responded by funding the Technology in Retrospect and Critical Events in Science (TRACES) project, which looked farther back in the history of various technological advances and found that many of them were made possible only because of successful earlier basic research. David Mowery and Nathan Rosenberg, in an ar-

---

20. Gibbons and Johnston (1974).

ticle summarizing and criticizing this debate, argued that focusing on either perception of demand or perception of a technological opportunity as the factor stimulating particular technological effort was pointless.[21] They advocated investing only in cases in which a scientific opportunity and a practical demand were present.

In many technologies, the early findings of Gibbons and Johnston continue to hold up—much of the science being drawn upon in the private sector is not new science. However, in some areas, the connections between university research and commercial application are relatively close: pharmaceuticals, certain other chemical technologies, various fields of electronics, and biotechnology. In these fields, inventors draw on science that is recent.

The nature of the interaction between application and the development of basic concepts was illuminated by a survey research project conducted in the mid-1980s. Industry executives in charge of R&D were asked about the importance of various bodies of basic and applied science for technical advance in their industry.[22] They were also asked about the relevance of current research in these scientific areas. Most respondents rated the relevance of a science much higher than the relevance of university research in that science. Evidence supports the interpretation that effective industry R&D in a specific field almost always required that the scientists and engineers working in industry must be trained in universities so they are familiar with the basic scientific understandings and techniques, but that in many cases new advances in that science were not exploited in industrial R&D. If the wetware (educational) and software (research) outputs of the university were separated, for most business, the output of wetware was what mattered.

For the most part, the industrial respondents tended to score most highly the relevance of university research in the engineering fields and in scientific fields such as materials science and computer science—fields in Pasteur's quadrant. Most of the respondents stated that university research in basic disciplines such as mathematics and physics was not particularly relevant to technical advance in their lines of business. This does not mean that basic research in the fundamental disciplines is not relevant to technical advance. It suggests that the results of basic research in fields such as mathematics and physics influence technical change

21. Mowery and Rosenberg (1979).
22. Klevorick and others (forthcoming).

indirectly, by improving and stimulating research in the more applied scientific and engineering disciplines.

Fields of technology differ significantly in the extent to which the science employed is of recent origin or relatively mature. One of the striking characteristics of technologies where advance is very rapid is that people doing applied research and development in those fields draw extensively on recent research findings. Thus a strong positive correlation exists between various measures of the rate of technical advance in a field and the significance of university research to technical advance in that field reported by the survey respondents.

Policy Implications

The confluence of forces that has led to a retreat from doctrines of science policy had been the norm in the half-century since World War II. Deep skepticism now is evident that ample funding of basic research at universities, with the areas of research determined largely by the scientists themselves, will inevitably create a wonderful cornucopia of useful things. A wide belief is that basic research funding needs to be more carefully targeted at national needs, or at least at different national needs. With the end of the cold war, the Department of Defense, which accounted for a large share of the burden of research funding in many fields, no longer can carry that load. For different reasons, the National Institutes of Health may no longer be able to pour increasing quantities of funding into basic research in the biomedical sciences. Until recently, only scattered political support could be found for the idea that the government should support research that is expressly designed to help firms produce goods for sale in private markets. Now the only question is which firms and how much. For all these reasons, public perceptions about science, technology, and economic growth have changed, and policy is changing with them. What does the economic understanding sketched out in this essay imply for policy?

While many scientists and academics rail against the idea that basic research priorities should be shaped by perceptions of national need, most basic research funding since World War II has come from mission-oriented government agencies, which decided what fields, and projects, to support. And much of academic basic research has been in the applied sciences and the engineering disciplines. What is new is not these facts, but the rhetoric in support of the targeting of basic research and the expressed desire by members of Congress for a greater say in determining the broad outlines of the research agenda. Judging from past experience, policymakers can set broad priorities and leave a powerful independent role for scientists in determining what is good science and what is not. Broad priorities are also consistent with considerable autonomy on the part of individual scientists and scientific groups in defining the details of their research agenda and carrying it out. The major risk from the newly articulated emphasis on national needs and the attempt to redefine what those needs are is that it will push too far in the direction of narrow targeting and quick payoffs. That is, both the Bohrs and the Pasteurs may be forced to become Edisons.

Academic research in the most fundamental of the scientific disciplines, such as various areas of physics and mathematics, never has accounted for a large share of the funding for academic research. A risk exists that these fields will be squeezed back even more and that cutbacks will be made in the basic research activities in biomedical science and in fields such as materials and computer science and the engineering disciplines. In all of these areas, important scientific and economic benefits have come from simultaneous, interrelated efforts to find applications for existing concepts and to discover new fundamental concepts. Cutbacks in the basic research activities that develop the basic concepts could impede the progress of science and reduce the rate at which valuable new applications are found.

No inherent danger can be discerned in moving toward an environment where economic and commercial opportunities are given more explicit weight in determining broad areas of national need and where national security and health carry less weight. This change poses little risk provided that it does not reduce the fraction of research that is focused on fundamental concepts and does not shorten the time horizon over which payoffs are measured. The best way to avoid such a shift would be to preserve the institutional arrangements for supporting research that have worked so well. Universities have offered an extremely effective environment for exploring basic concepts and pursuing distant payoffs. A shift toward commercial and economic objectives should be accomplished by changing the emphasis in university research, not by pushing that research into the private sector. There must continue to be a place in the university for modern day Pasteurs.

The returns from this attempt to adjust priorities will be larger if it is accompanied by two complementary developments. One is a change in orientation of advanced training programs in the sciences and engineering disciplines. They should move toward training people for work in the private sector and away from the presumption that Ph.D.s, or at least good ones, get recycled into academia. A great step toward this goal could be taken merely by changing the attitudes and expectations that permeate the graduate faculty. Changing attitudes and expectations will not be easy, but the alternative is to stand by while the number and quality of people getting advanced training in the sciences declines. In an era of rapidly unfolding technological opportunities, cutting back on advanced training in science would be unwise.

If university research and graduate training are to be oriented more toward the needs of industry, the mechanisms for interaction between university and industry scientists and engineers should be widened and strengthened. Universities and companies might strive for a significant increase in the extent to which industry scientists spend periods of time in academia, and academic scientists in industry. These exchanges might even be supported by government funds. Instead of giving money directly to firms to do research on specific topics, the government might fund competitive fellowships that encourage these exchanges. The government might also explicitly subsidize the training of students who will go to work in the private sector. By taking these steps, the government could subsidize the inputs that go into private sector research instead of contracting with firms for specific research outputs. Market demands and market perceptions of opportunity thus would continue to be the primary force that allocates resources among specific research projects in the private sector. The pork-barrel politics that can arise when the government writes checks to business firms can be avoided.

Establishing property rights on the output from scientific research is generally not a good practice. This is true whether that research is directed at practical problems facing the military, health professionals, or business firms. Important efficiency advantages are evident in a system in which the government subsidizes the production of fundamental concepts and insights and gives them away free. The Bayh-Dole act of 1980 marked a major retreat from the principle that knowledge subsidized by the government should circulate freely, and the continuing argument about issues such as whether gene fragments ought to be patentable reflects strong pressures to move even further in this direction. Even as property rights are strengthened on the applications end of the software spectrum, private property rights should not be set for bodies of knowledge and techniques that have wide and nonrivalrous applications, particularly when many of these applications are in further research and development. A renewed attention to the needs of industry does not have to be associated with a major change in the intellectual property rights regime. No reason exists to treat science as private instead of public knowledge.

World War II produced a new set of principles about the role of the federal government in support of science. The arguments presented in Vannevar Bush's report captured some of these principles. The major support that the Defense Department and the National Institutes of Health provided for mission-oriented basic research reflected others. This new understanding encompassed the traditional principle that private funds should be the main support for commercial applications of science, the kind of activity undertaken by Thomas Edison. To this were added a new set of principles about science: Government funds should be used to finance the search for new fundamental concepts and insights. Good science can emerge from an internal dynamic that drives scientists to uncover new concepts and refine old ones, the kind of dynamic that motivated Niels Bohr. Good science can also arise from the efforts of someone such as Louis Pasteur, someone who tries to respond to external demands for specific kinds of applications and to understand the applications developed by others. Within a framework of priorities set by policymakers, the scientific community should have the principal voice in funding and evaluating science in all of its forms. And scientific knowledge should be a public good.

These principles are as relevant today as they were then. The details of science and technology policy should be adjusted in response to changing circumstances. Attention should be shifted to new applications that will feed into—and follow from—the search for fundamental scientific concepts. But the principles should not change.

References

Abramovitz, M. 1956. "Resource and Output Trends in the United States since 1870." *A. E. R. Papers and Procedures* 46 (May): 5–23.

———. 1986. "Catching Up, Forging Ahead, and Falling Behind." *Journal of Economic History* 46 (2): 385–406.

Arrow, K. J. 1962. "Economic Welfare and the Allocation of Resources for Invention." In *Rate and Direction of Inventive Activity*, 609–25. Cambridge, Mass.: National Bureau of Economic Research and Princeton University Press.

Baily, M. N., and A. K. Chakrabarti. 1988. *Innovation and the Productivity Crisis.* Brookings.

Barro, R., and X. Sala i Martin. 1991. "Economic Growth across States and Regions." *Brookings Papers on Economic Activity*: 107–58.

Cornes, R., and T. Sandler. 1986. *The Theory of Externalities, Public Goods, and Club Goods.* Cambridge University Press.

Gibbons, M., and R. Johnston. 1974. "The Roles of Science in Technological Innovation." *Research Policy* (3).

Griliches, Z. 1958. "Research Costs and Social Returns: Hybrid Corn and Related Innovations." *Journal of Political Economy* 66 (October): 419–31.

———. 1992. "The Search for R&D Spillovers." *Scandinavian Journal of Economics* 94 (Supplement): S29–47.

———. 1994. "Productivity, R&D, and the Data Constraint." *American Economic Review* 84 (March): 1–23.

Jorgenson, D. W., F. Gollop, and B. Fraumeni. 1987. *Productivity and U.S. Economic Growth.* Harvard University Press.

Kay, Lily E. 1993. *The Molecular Visions of Life.* Oxford University Press.

Kevles, D. J. 1979. *The Physicists.* Vantage Books.

Klevorick, A. K., and others. "On the Sources and Significance of Interindustry Differences in Technological Opportunities." *Research Policy* 24:185–207.

Kline, S. J., and N. Rosenberg. 1986. "An Overview of Innovation." In *The Positive Sum Strategy: Harnessing Technology for Economic Growth,* edited by R. Landau and N. Rosenberg. Washington, D.C.: National Academy Press.

Levin, R. C., and others. 1987. "Appropriating the Returns from Industrial R&D." *Brookings Papers on Economic Activity*: 783–820.

Mansfield, E., and others. 1977. *The Production and Application of New Industrial Technology.* New York: W. W. Norton.

Mowery, D. C., and N. Rosenberg. 1979. "The Influence of Market Demand on Innovation: A Critical Review of Some Empirical Studies." *Research Policy* 8 (April): 103–53.

Nelson, R. R., and G. Wright. 1992. "The Rise and Fall of American Technological Leadership: The Postwar Era in Historical Perspective." *Journal of Economic Literature* 30: 1931–64.

Romer, P. M. 1990. "Are Nonconvexities Important for Understanding Growth?" *American Economic Review* (80): 97–103.

———. 1990. "Endogenous Technological Change." *Journal of Political Economy* 98: S71–S102.

———. 1994. "The Origins of Endogenous Growth." *Journal of Economic Perspectives* 8: 3–23.

Rosenberg, N. 1982. "How Exogenous Is Science?" In *Inside the Black Box: Technology and Economics,* edited by Nathan Rosenberg. Cambridge University Press.

————. 1994. "Neglected Uncertainties in the Process of Technological Change." Stanford University.

Rosenberg, N., and R. R. Nelson. 1994. "American Universities and Technical Advance in Industry." *Research Policy* 23: 323–48.

Solow, R. 1957. "Technical Change and the Aggregate Production Function." *Review of Economics and Statistics* 39: 312–20.

Starrett, D. 1988. *Foundations of Public Economics*. Cambridge University Press.

Stiglitz, J. E. 1988. *Economics of the Public Sector.* W. W. Norton.

Stokes, Donald E. Forthcoming. *Pasteur's Quadrant: Basic Science and Technological Innovation*. Brookings.

CHAPTER 4

Contributions of R&D to Economic Growth

Michael J. Boskin and Lawrence J. Lau

Hᴉsᴛᴏʀʏ is replete with examples, even epochs, of new technology generating faster economic growth: from railroads to electricity to jet planes. Many discussions of the evolution and growth of modern economies focus on new products, from pharmaceuticals to fax machines. While numerous impressive case studies and historical analyses exist, by themselves they are not able to establish the importance of research and development (R&D) to overall economic growth. To do that, simultaneous case studies would be needed on much of the economy. Is most of the new technology derived from formal and informal research and development? Is technological advance the primary source of economic growth? Economists have explored the contribution of research and development to economic growth with great vigor for several decades. Several strands of thought have been developed, and analytical arguments and empirical evidence mustered, to support the proposition that research and development strongly contributes to economic growth and that private incentives are insufficient to produce the socially optimal investments in research and development. The two main approaches are studies, theoretical and empirical, of rates of return to R&D expenditures and econometric and other studies of the sources of aggregate economic growth.

Theoretical analysis generally has focused on the potential inability of private investors to appropriate fully the returns from investment in R&D. The argument most easily applies to basic physics research. While some aspects of the knowledge may be privately appropriable, much of the knowledge will be a public good, and others, including potential competitors, cannot be excluded from its use. Or at least, the patent laws may be an ineffective way of assuring private appropriability. Or alternatively, even if the R&D does not produce a classic pure public good, there may be sufficient spillover benefits that individual private investors could not appropriate; left to its own devices, the private sector would systematically underinvest in such activities. Two-thirds of R&D is D, and not all R has these characteristics.

Economists have been most comfortable with assuming the lack of full appropriability in basic research; less comfortable in applied research and development. And this is where the dividing line is often drawn in academic, and political, arguments over industrial policy and government subsidization of research and development. But the dual criteria for a case to be made for government subsidization of private investors' being unable to appropriate fully and the expected social benefits' exceeding the expected social costs do not easily fit into any simple taxonomy. Legitimate cases may arise further down the road toward applied research or even development. An intelligent judgment of the merits of the case would require technical scientific, as well as economic, input. The government is heavily in the R&D business: It conducts much R&D itself in government laboratories; it helps finance much private R&D, directly and through tax credits; and it purchases the output of much R&D. So while serious analytical argument exists that a totally laissez-faire system would systematically underinvest in R&D, that may not be a reality given the current level of government involvement.

An argument that is related theoretically stresses other types of spillovers. For example, learning-by-doing leads to some form of increasing returns—more investment opens up new frontiers and opportunities.

These types of analytical arguments have their counterpart in the important empirical literature attempting to estimate social rates of return to R&D. Most of the econometric work is conducted on industry-level data. For example, Zvi Griliches reports regressions on three-digit Standard Industrial Classification (SIC) level data for different postwar periods of the growth of total factor productivity (the growth in output unaccounted for by the growth in inputs) on the rate of investment in R&D (as measured by the R&D-to-sales ratio).[1] The estimated coefficient is interpreted as the excess gross rate of return to R&D and is on the order of 30 percent. The question immediately becomes, accepting the results at face value, why does such a large excess return persist? One is then thrown back on appropriability or a related argument.

Note, however, that even accepting the finding of a large divergence

1. Griliches (1994).

between the social and private rate of return to R&D, at the ex post levels of R&D investment observed, this does not imply that increased R&D investment would have a large impact on aggregate growth. There are several reasons. First, evidence of a local divergence in the social and private rates of return says little or nothing about what would happen to that differential if R&D were to expand substantially. It may well be that large excess social returns do exist in many industries, but that a very small expansion would exhaust almost all of the excess social returns from greater investment in R&D and that the overall contribution to aggregate economic growth would be small. Still, the appropriability and spillover arguments do have a common-sense ring and some empirical support. They also have become important in analytical formulations in modern growth theory. For example, the generation of knowledge in the innovation process may increase the knowledge base, which in turn increases the returns to human capital in future research, leading to an increase in the growth rate.[2] R&D expenditures may produce multiple equilibria via increasing returns, or spillovers, and social benefits may be derived from getting above a certain threshold that would change private incentives.

Potentially offsetting this notion of underinvestment in R&D is the allegation that much of the R&D helps create, at least temporarily, monopoly positions, in which output is restricted, prices raised, and monopoly rents earned. In moving beyond the static notion, with a nod toward Joseph Schumpeter, R&D investment may generate a new technology that replaces the old one, transferring the monopoly rents to another firm's shareholders. Going still further, in industries considered particularly R&D intensive, and where allegations of monopoly practices abound—and are often deemed desirable as a way to generate R&D protected by the patent laws—such as pharmaceuticals and microelectronics, a great deal of reinvestment of the profits into additional R&D occurs. This immediately raises the question of whether the external capital market would be able to finance the R&D, or whether there is something peculiar to the nature of the risk or information that requires great reliance on internal financing out of these monopoly profits. If the latter is true, it creates a situation in which an accrued R&D asset will, in expected value terms, generate future benefits, but the value will not be included in traditional national income accounts. It might be observed in the stock market's evaluation of such firms (although the stock mar-

2. Romer (1990).

ket's evaluation of R&D appears to have fallen).[3] In any event, this could be an important omitted item in the national balance sheets.

The microeconomic analysis and econometric estimates arguing that a systematic underinvestment in R&D exists and that aggregate growth could be raised measurably by expanding R&D investment are interesting and important. However, no convincing case could be made that R&D investment would be enough to be a major contributor to higher aggregate growth. More R&D may be desirable for many industries, but current institutions and practices—from patent law and other intellectual property protections to tax rules and government regulation—may limit research's contribution to growth. Expansion of R&D in such industries also could rapidly use up any excess social returns with little extra contribution to aggregate output beyond the optimal decisions made by private businesses and investors or increased R&D could merely result in a reallocation of monopoly rents. The purpose here is not to dismiss this literature, only to point out some of its limitations. Were it tied to strong statistical evidence that the impact of R&D at the aggregate level is large and that its rate of return substantially exceeds the rate of return on alternative types of capital, a stronger case could be made. For that reason, the remainder of this paper focuses on the second approach to analyzing the contributions of R&D to economic growth—namely, the sources of aggregate economic growth approach.

The analysis of the sources of economic growth generally starts with an aggregate production function relating some measure of real output to various factors of production, such as traditional tangible capital and labor. For the United States in the late nineteenth and early twentieth century, statistical studies demonstrate that most of the growth in aggregate output can be explained by the increase in capital and labor.[4] But beginning in the late 1950s, empirical studies of the American economy for the period starting in the late 1920s discovered that the growth in conventionally measured inputs explained only a modest fraction of the observed growth and output. Moe Abramovitz dubbed this "a measure of our ignorance."[5]

The use of the residual as a method for measuring technological advance and the assumption that most of the technological advance was caused by "formal and informal R&D investments by individuals, firms,

---

3. See Hall and Hall (1993).
4. Denison (1962); Kuznets (1971).
5. Abramovitz (1956).

and governments, and the largely unmeasured contributions of science and other spillovers" gave a strong boost to the case that R&D was immensely important to long-run economic growth.[6] Considerable technical change was brought about, and assumably most or all resulted from R&D, which was enormously important to explaining the growth of aggregate output.

Numerous studies tried to explain this so-called residual. Attempts were made to measure quality change in capital and labor, and much of what was called technical change was attributed to improvements in the quality of capital and labor, and hence mismeasurement of the growth of traditional inputs. This begs the question of what produces the quality change, and, for the purposes of this paper, how much is the result of formal and informal R&D investment. Next, the notion of capital was expanded to include R&D capital. Measures of the R&D capital stock comparable to the traditional tangible capital stock were developed. The effects of R&D capital on growth were small; attempts to account for the residual by increasing returns to scale were limited.[7] Perhaps the measurement problems were too severe, but these various micro conjectures did not add up to a large part of the aggregate story.[8] The reason could be that other important factors were evident in the residual—organizational efficiency, improvements in or impediments to the efficiency of market mechanisms, and so on. Or that measuring the concepts involved was too difficult, or that the data were constructed in a manner in which the benefits from R&D spillovers or increasing returns could not show up.

Technical Change and the Residual in Growth Accounting

The investigation of the effects of R&D on economic growth begins by focusing on the residual in growth accounting. Some authors assume the residual measures technical change, which in turn is generated by R&D (even if R&D is only part of the story). Sometimes the assumption is that the larger the residual, the greater the contribution of R&D. Two approaches to the measurement of the residual, or technical process, will be considered—growth accounting and econometric estimation.

6. Griliches (1994).
7. See, for example, Griliches (1988).
8. See, especially, Griliches (1994).

The point of departure of most studies of productivity and growth accounting is the aggregate production function:

(4-1) $$Y_t = F(K_t, L_t, t),$$

where Y_t, K_t, and L_t are the quantities of aggregate real output, physical capital, and labor, respectively, at time t, and t is an index of chronological time. The rate of growth of output can be expressed in the familiar equation of growth accounting by taking natural logarithms of both sides of equation 4-1 and differentiating it with respect to t:

(4-2)
$$\frac{d\ln Y_t}{dt} = \frac{\partial \ln F}{\partial \ln K}(K_t, L_t, t)\frac{d\ln K_t}{dt} + \frac{\partial \ln F}{\partial \ln L}(K_t, L_t, t)\frac{d\ln L_t}{dt}$$
$$+ \frac{\partial \ln F}{\partial t}(K_t, L_t, t),$$

where $\dfrac{d\ln Y_t}{dt}$, $\dfrac{d\ln K_t}{dt}$, and $\dfrac{d\ln L_t}{dt}$ are the instantaneous proportional rates of change of the quantities of real output, capital, and labor, respectively, at time t; $\dfrac{\partial \ln F}{\partial \ln K}$ and $\dfrac{\partial \ln F}{\partial \ln L}$ are the elasticities of real output with respect to capital and labor, respectively, at time t; and $\dfrac{\partial \ln F}{\partial t}$ is the instantaneous rate of growth of output holding the inputs constant, or equivalently, the rate of technical progress.

The three terms on the right-hand side of equation 4-2 may be identified as the contribution of capital, labor, and technical progress, respectively, to the growth in output.[9]

The purpose of growth accounting is to determine from the empirical data how much of the change in real output between, say, $t = 0$ and $t = T$ can be attributed to changes in the inputs, capital and labor, and technology, respectively. The first term on the right-hand side of equation 4-2 represents the instantaneous contribution of the growth in capital input to the growth in real output. Note that the contribution of capital is the product of the production elasticity of capital and the rate of growth of capital. If the rate of growth of capital is low, then the contribution of capital may be low even with a high production elasticity of capital, and vice versa. Similarly, the second term represents the instan-

9. In almost all such formulations, technical progress is taken to be exogenous.

taneous contribution of the growth in labor input, and the third term represents the instantaneous contribution of technical progress. Together, the three terms add up to the instantaneous rate of growth of aggregate real output at time t.

Equation 4-2 is the fundamental equation of growth accounting. Its function is twofold. First, it can be used as the basis of growth accounting, a decomposition of the growth in real output into the growth in its three separate sources—capital, labor, and technology—each represented by one of the three separate terms. Second, equation 4-2 may be rewritten as

(4-3)
$$\frac{\partial lnF}{\partial t}(Kt,Lt,t) =$$

$$\frac{dlnYt}{dt} - \left[\frac{\partial lnF}{\partial lnK}(Kt,Lt,t)\frac{dlnKt}{dt} + \frac{\partial lnF}{\partial lnL}(Kt,Lt,t)\frac{dlnLt}{dt}\right],$$

which can be used to obtain an estimate of technical progress (or the growth in total factor productivity), given the values of the different quantities on its right-hand side. However, in general, not every quantity on the right-hand side of equation 4-2 or equation 4-3 can be directly observed. Only the rates of growth of aggregate real output, capital, and labor are directly measurable as data. The elasticities of output with respect to capital and labor must be separately estimated, often requiring additional assumptions. Moreover, the instantaneous rate of technical progress, $\frac{\partial lnF}{\partial t}$ (K_t,L_t,t), depends on K_t and L_t as well as t, except in the case of neutral technical progress.[10] One may be tempted to sum up the instantaneous rate of technical progress in each period to arrive at an estimate of the technical progress over T periods as

(4-4)
$$\int_0^T \frac{\partial lnF}{\partial t}(K_t,L_t,t)dt.$$

However, equation 4-4 can be rigorously justified only if either (1) technical progress $= \left(\frac{\partial lnF(t)}{dt}\right)$ is neutral; that is,

---

10. More specifically, except in the case that the production function takes the form $Yt = A(t) f(Kt, Lt)$. Neutrality of technical progress implies a somewhat weaker condition: $Yt = F(f(Kt, Lt), t)$, where $f(\cdot)$ is a real-valued function of capital and labor.

(4-5) $$1nF(K_t,L_t,t) = 1nA(t) + 1nf(K_t,L_t),$$

so that $\dfrac{\partial 1nF}{\partial t}$, the instantaneous rate of technical progress, is independent

of K_t and L_t; or (2) capital and labor remain constant over time, so that

(4-6) $$\int_0^T \frac{\partial 1nF}{\partial t}(K_t,L_t,t)dt = \int_0^T \frac{\partial 1nF}{\partial t}(K_0,L_0,t)dt$$

$$= 1nF(K_0,L_0,T) - 1nF(K_0,L_0,0),$$

which is precisely the logarithmic growth in real output holding inputs constant. However, if technical progress is not neutral, then to the extent that K_t and L_t change over time, the rate of technical progress over many periods cannot be simply cumulated from one period to the next.

Thus, in general, when equation 4-3 is used to measure technical progress over time, three basic hypotheses are maintained: constant returns to scale, neutrality of technical progress, and profit maximization with competitive output and factor markets. Under profit maximization with competitive markets, the elasticity of output with respect to labor is equal to the share of labor cost in the value of total output. As a result, the labor share, which is directly observable, can be used as an estimate for the production elasticity of labor. Constant returns to scale in production implies that the sum of the production elasticities of capital and labor is equal to unity, so that the elasticity of capital can be readily estimated as one minus the elasticity of labor when the latter is known. The constant returns to scale assumption is crucial to the estimation of the production elasticity of capital because, while, in principle, it can be independently estimated by the share of capital cost in the value of total output under the assumption of profit maximization, in practice, data on the rental price, or user cost, of capital are seldom available or reliable. Finally, neutrality of technical progress implies that $\partial 1nF/\partial t$ is independent of capital and labor. This assumption is used to justify the cumulation of successive estimates of technical progress over time even as K_t and L_t change their values.

However, even equations 4-2 and 4-3 cannot be directly implemented because data are not available in continuous time. For example, data on aggregate real output and inputs are typically available as the total flow between, say, time t and time $t+1$. Technical progress between time t and time $t+1$ is given by the integral of equation 4-3 between time t and time $t+1$:[11]

11. This is sometimes referred to as the Divisia index.

$$\int_t^{t+1} \frac{\partial \ln F}{\partial t}(K_t, L_t, t)dt = \int_t^{t+1} \frac{d \ln Y_t}{dt}dt$$

(4-7)
$$- \int_t^{t+1} \frac{\partial \ln F}{\partial \ln K}(K_t, L_t, t)\frac{d \ln K_t}{dt}dt$$

$$- \int_t^{t+1} \frac{\partial \ln F}{\partial \ln L}(K_t, L_t, t)\frac{d \ln L_t}{dt}dt.$$

However, because data are not available in continuous time, a discrete approximation of the integrals in equation 4-5 is usually made. One such approximation is the Tornqvist index, given by[12]

(4-8) $$\int_t^{t+1} \frac{\partial \ln F}{\partial t}(K_t, L_t, t)dt \cong \ln Y_{t+1} - \ln Y_t$$

$$- \frac{(s_{K(t+1)} + s_{Kt})}{2}(\ln K_{t+1} - \ln K_t) - \frac{(s_{L(t+1)} + s_{Lt})}{2}(\ln L_{t+1}) - \ln L_t),$$

where s_{Kt} and s_{Lt} are the values of the production elasticities of capital and labor, respectively, at time t. Other approximations are possible.

A leading alternative to using equation 4-2 or 4-3 (or the derivative equations 4-7 or 4-8) to measure technical progress or to account for growth is the direct econometric estimation of the aggregate production function, $Y_t = F(K_t, L_t, t)$, from data on aggregate real output and capital and labor inputs. Such direct estimation of the production function does not, in principle, require any assumption beyond that of the functional form. In particular, it does not require the traditionally maintained assumptions of constant returns to scale, neutrality of technical progress, and profit maximization with competitive output and input markets. In practice, however, the hypothesis of constant returns to scale is often maintained, especially when data of a single country are analyzed, because it reduces the number of independent parameters to be estimated and thereby mitigates the possible multicollinearity among the data on capital and labor inputs and time.[13]

It is tempting to use the estimated production function to derive estimates of the production elasticities of capital and labor, and then to use

12. For a discussion of the Tornqvist index, see, for example, Jorgenson, Gollop, and Fraumeni (1987).
13. However, the problem of multicollinearity may also be partially mitigated by pooling time-series data across countries. See the discussions in Lau and Yotopoulos (1989) and Boskin and Lau (1990).

these estimated elasticities to implement equation 4-2 or 4-3. However, such a procedure generally results in an estimate of technical progress that is statistically defective. An example will make this point clear. Suppose the aggregate production function is assumed to have the Cobb-Douglas form, and the following equation is estimated:

$$(4\text{-}9) \qquad 1nY_t = 1nA_0 + \alpha 1nk_t + \beta 1nL_t.$$

Let $\hat{\alpha}$ and $\hat{\beta}$ be the estimated production elasticities of capital and labor, respectively. Substituting these estimated elasticities into, say, equation 4-8, the right-hand side may be immediately recognized as simply the first difference of the residuals from the estimated equation 4-9, which, under the usual assumptions, have zero expected values. (If ordinary least-squares is used to estimate equation 4-9, the residuals will sum to zero over the sample.) Thus any estimate of technical progress obtained this way, whether for a single period or over the whole sample period, is meaningless.[14] The appropriate way to obtain an estimate of technical progress if an aggregate production function is to be directly estimated is to include a term (or terms) for the time trend in the aggregate production function to capture the effect of technical progress.[15] Once this is done, however, equation 4-2 or 4-3 becomes redundant, because all the information on technical progress is contained in the estimated aggregate production function. In general, if the aggregate production function is specified correctly (with or without the time trend term(s)), the residuals cannot be interpreted as estimates of technical progress; and if the aggregate production function is specified incorrectly (say, by omitting the time trend term(s) when they should be included), the estimated parameters, and hence any functions of these parameters, are biased and unlikely to be useful.

Finally, what are some of the pitfalls of maintaining the traditional assumptions of constant returns to scale, neutrality of technical progress, and profit maximization with competitive output and input markets in the measurement of technical progress and growth accounting? First, for an economy in which aggregate real output and inputs are all growing

---

14. If the time trend term belongs in the aggregate production function in equation 4-9, then not including it in the estimation will in general lead to biased estimates of the production elasticities, unless the capital and labor input series are orthogonal to the time trend, an unlikely situation.

15. More than one term could be included to allow technical progress to be nonlinear (nonexponential) over time.

over time, identifying separately the effects of returns to scale and technical progress—either one can be used as a substitute explanation for the other—is difficult. Thus, to the extent that there are increasing returns to scale, maintaining the hypothesis of constant returns to scale results in an overestimate of technical progress; and to the extent that there are decreasing returns to scale, maintaining the hypothesis results in an underestimate. An examination of equation 4-3 provides the clearest example. If there are increasing returns to scale, the assumption of constant returns to scale implies that the sum of the production elasticities is underestimated and hence at least one but possibly both of the production elasticities are underestimated, with the consequence that the rate of technical progress will be overestimated. A further consequence is that the contributions of the capital and labor inputs to economic growth will also be underestimated. The reverse is true if there are decreasing returns to scale.

Second, if technical progress is non-neutral, then the rate of technical progress at time t will vary depending on the quantities of capital and labor inputs at time t. Moreover, technical progress over many periods cannot be simply cumulated as the sum of the technical progress that has occurred over the individual periods, nor can they be simply averaged.

Third, the hypothesis of profit maximization (or cost minimization) with competitive output and input markets implies the equality of the production elasticities of capital and labor with the respective factor shares. To the extent that constraints are placed on instantaneous adjustments of inputs to their desired levels or monopolistic or monopsonistic influences are felt in the output and input markets, the production elasticities may deviate from the factor shares and an estimate of technical progress obtained from equation 4-3 or 4-8, using factor shares as estimates of elasticities, will be subject to biases. In addition, the estimated contributions of capital and labor inputs to economic growth will also be subject to biases. For example, if the output market is monopolistic, then the factor shares are likely to underestimate the production elasticities, causing in turn an overestimate of technical progress if an equation such as 4-3 or 4-8 is implemented. If an input market is monopsonistic, then the reverse is likely to be true.

The direct econometric estimation of the aggregate production function is in principle free of all these biases because it does not require the traditional assumptions as part of the maintained hypothesis. However, it will be subject to the same biases if any of the traditional assumptions is kept as part of the maintained hypothesis of the econometric model.

Identifiability of the Residual from Aggregate Input-Output Data

Technical progress is usually identified with the residual in traditional growth accounting, or in econometric analysis, with the effect of the time trend term. A number of difficulties arise with the empirical identification and estimation of technical progress from data on a single country in the absence of assumptions on the nature of the scale effects and technical change.

Difficulties in the Empirical Measurement of Technical Progress

THE CONFOUNDING OF ECONOMIES OF SCALE AND TECHNICAL PROGRESS. The effects of technical progress and scale cannot be separately identified with aggregate time-series data from a single country in which output and inputs have both been growing. In this case, the growth in output can be attributed to either economies of scale or to technical progress. Empirically, one cannot distinguish between the two sources of growth because only a single observation can be made at any particular time. As a result, considerable indeterminateness exists in the measured rate of technical progress, which is sensitive to the maintained hypotheses on the nature of the scale effect and technical progress.

These ideas are illustrated graphically with a one-output and one-input (capital) case. In figure 4-1, four observations of input-output combinations were plotted, one for each period. In figure 4-2, the restriction of constant returns to scale was imposed, so that the production function is a straight line through the origin. There is no technical progress. In figure 4-3, for the same four observations, decreasing returns to scale production functions pass through each of the four points. There is technical progress from period to period.

The problem of underidentification can be solved by pooling time-series data for different countries together, so that, at any given time, different countries operate at different scales and the same scale is also observed at different times.[16]

16. To justify the pooling of data across countries, some assumptions of similarity must be made, which is done in the metaproduction function approach. However, these assumptions of similarity—the maintained hypotheses of the metaproduction function approach—can be statistically tested and cannot be rejected in almost all of the applications made using the metaproduction function approach.

Figure 4-1. *Economies of Scale and Technical Progress: Original Observations of Input-Output Combinations*

Output (Y)

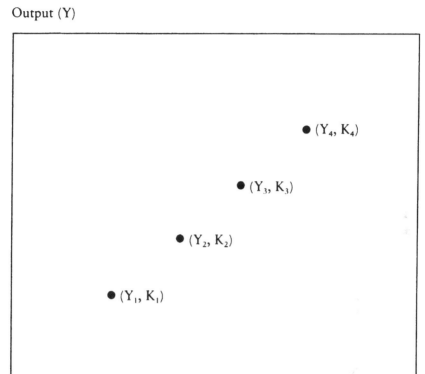

Capital (K)

THE UNDERIDENTIFICATION OF THE BIASES OF SCALE EFFECTS AND TECHNICAL PROGRESS. The assumptions on the nature of the scale effects and technical progress also affect the measurement. For example, in figure 4-4, again for the same four observations, neutrality of technical progress is assumed, resulting in estimates of technical progress that are independent of the period of measurement (or cumulative over successive periods). However, in figure 4-5, in which neutrality is not assumed, different conclusions are reached on the magnitudes of technical progress.

Because separate identification of the technical progress and scale effects for an economy in which the output and all the inputs are growing over time is impossible, a fortiori, the identification of the biases of technical progress and scale effects is also impossible. In general, with observations from only a single country, identifying the biases is not possible. With the metaproduction function approach, identifying the

Figure 4-2. *Economies of Scale and Technical Progress:*[a] *Constant Returns to Scale Assumed*

Output (Y)

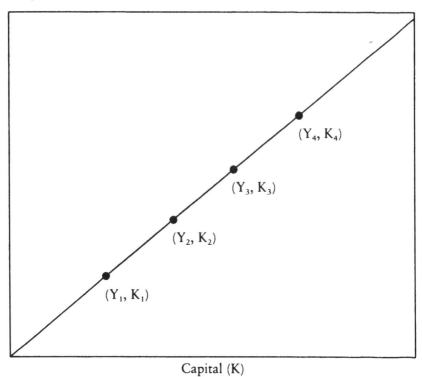

Capital (K)

a. There is zero technical progress.

rate of technical progress, the degree of economies of scale and their biases is possible.

The Metaproduction Function Approach

To enable the pooling of data, assume that the aggregate production functions of the different countries are identical, in terms not of measured outputs and inputs but of efficiency-equivalent units of outputs and inputs. An additional advantage of the metaproduction function approach is the possibility of identifying the relative efficiencies or qualities of the outputs and inputs of the different nations and their changes over time. For example, using this approach, annual estimates of the efficiency-

Figure 4-3. *Economies of Scale and Technical Progress:*[a] *Decreasing Returns to Scale Assumed*

Output (Y)

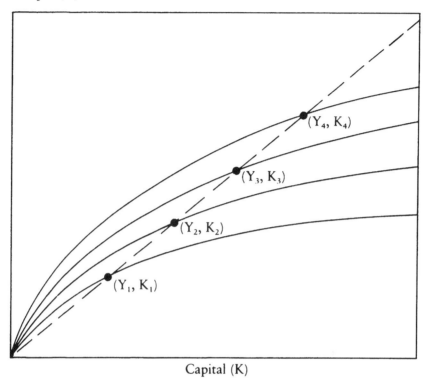

Capital (K)

a. There is technical progress from period to period.

equivalent conversion factors between Japanese and United States labor in the postwar period can be obtained.

The Degree of Returns to Scale

The degree of (local) returns to scale is a critical parameter in the empirical measurement of the residual because they are generally inversely related. For a given data set, the higher the assumed or estimated degree of returns to scale, the lower is the estimated contribution of the residual. Traditionally, constant returns to scale is the standard assumption, partly based on the superficially plausible grounds that replications should always be possible. However, the replicability argument is valid only if all inputs are included and are variable. Returns to scale may be

Figure 4-4. *Economies of Scale and Technical Progress:*[a] *Neutrality of Technical Progress Assumed*

Output (Y)

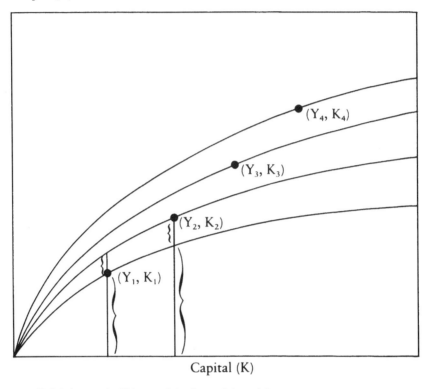

Capital (K)

a. Technical progress is additive (cumulative) from period to period.

observed to be decreasing if important factor inputs are omitted from the aggregate production function. For example, if land (or, more generally, the natural endowment of resources) is not explicitly included as a factor of production in the empirical analysis, then even if returns to scale were constant or increasing with respect to capital, labor, and land taken together, they would not necessarily be constant or increasing with respect to only capital and labor, holding land constant. Because land is fixed over time, to the extent that it is an important factor of production, one should not expect that simply doubling capital and labor but holding land constant will double real output. The same argument applies to the natural resource base, which may be declining in some countries. Other potentially important factors of production that are frequently omitted include human capital and R&D or knowledge capital, both of which

Figure 4-5. *Economies of Scale and Technical Progress:*[a] *Neutrality of Technical Progress Not Assumed*

Output (Y)

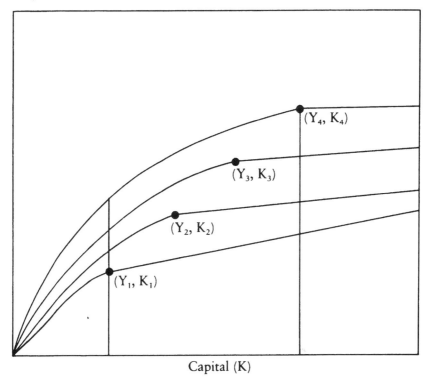

Capital (K)

a. Technical progress depends on the point of measurement; the bias cannot be uniquely identified.

have been growing over time. This line of argument suggests that returns to scale in the measured variable inputs may be decreasing.

However, plausible arguments also exist as to why returns to scale may be expected to be increasing. One is based on the well-established existence of increasing returns to scale in production at the microeconomic level for many industries; for example, automobile manufacturing, electricity generation, and petrochemicals. The interesting question is whether such economies of scale will manifest themselves at the macroeconomic level, in the sense that if the aggregate inputs are increased in the same proportion, aggregate output will increase more than proportionately. For one thing, if returns were increasing over some range of inputs at the plant level, one would expect, in equilibrium, that almost all plants would be at the efficient scale and that these efficient-scale plants can be replicated. Thus an economy sufficiently large relative to the sizes of minimum efficient-scale plants would exhibit constant returns to scale at the aggregate level.

A second argument is based on the technology for R&D (or the production of knowledge) that is characterized by a high fixed cost. However, once knowledge is produced, it can be distributed to and utilized by almost everyone at very low marginal costs. Thus there will be significant increasing returns. However, the effect of knowledge on production, even if freely available, can be exhausted, despite increasing returns to scale in the production and dissemination of knowledge. (This is related to the problem of a proper metric for the measurement of knowledge in the production function—while increasing returns are evident in knowledge production itself, decreasing returns ultimately will result in the utilization of knowledge capital in the production of aggregate output.)

A third argument for increasing returns to scale is based on the existence of coordination externalities. If all industries expand simultaneously, then they can be mutually sustaining; however, if industries attempt to expand individually, they will often fail. Thus there are economies of scale. But the scale economies are essentially realized by moving from one equilibrium to another in a multiple-equilibria world. Once the desirable equilibrium is reached (and a coordination mechanism is established), no expectation exists that such economies will continue to be available.[17] Empirically, the movement from one equilibrium to another will show up as a one-time improvement in productive efficiency, not economies of scale.

A fourth argument is based on network externalities. A typical example is the telephone system. If only one household has a telephone, it is useless—there is no one to call and no one will be calling. However, as the number of households with telephones increases, the utility of having a telephone increases for every telephone subscriber. There are thus increasing returns. However, these increasing returns will eventually be offset by congestion costs and other service failures as the network becomes large. Given the level of the technology, there is an efficient-scale network. Once this scale is attained, the national telephone system as a whole will essentially exhibit constant returns to scale (the replication argument also applies here).[18]

17. The mechanism does not need to be formal; it can take the form of information exchange, signaling, or more generally some formal or informal way of attaining common expectations about the future.
18. Education is another area in which there may be network-type externalities (for

The practical consequences of the existence of significant economies of scale in all inputs taken together may not be very large in the short and intermediate run because some of the inputs, such as human capital, cannot be quickly expanded. A long gestation period is required. Other inputs, such as land and natural resources, may be fixed or declining in supply. Physical capital as well as R&D capital are the only inputs that can be readily expanded in an intermediate time horizon.

The Effect of Learning-by-Doing

Learning-by-doing, a concept created by K. J. Arrow, is essentially a microeconomic phenomenon applying at the enterprise or the plant level.[19] While the cumulative production or cumulative investment experience in a single industry offers certain lessons, learning largely occurs based on the aggregate experience (although many agents can be simultaneously learning in different industries and enterprises). Learning effects at the microeconomic level also create local monopolies, which may be durable and persistent. The effect of learning-by-doing on output and efficiency at the aggregate level depends on the interaction between the efficiency gain from learning and the efficiency loss resulting from the monopolistic market power acquired. In any case, learning-by-doing will be manifested at the aggregate level mostly as part of the residual—over time, even with the same inputs, greater outputs will be produced.

Sometimes learning effects may be undone or nullified because of rules and regulations. For example, a team of workers that has gained greatly in efficiency in the manufacture of a product may be displaced by another team with more seniority but less experience on that particular product because of seniority rules for layoffs or furloughs.

Estimating the Contribution of the Residual to U.S. Economic Growth

How large is the residual? What are the results of the growth accounting exercises conducted by the various authors for the United States using the conventional approach? In table 4-1, the results are presented of

---

example, the degree of general literacy is important). Here the limitation can come from communication and transactional costs.

19. Arrow (1962).

Table 4-1. Comparison of Growth Accounting by Authors and Time Periods

Percent

Authors; time	uK^a	K/K^b Equipment	Inventories	Quality	wL^c	L/Ld Employment	Hours (quantity)	Hours (quality)	Education	Women exp. and util.	Age-sex component	Scale	λ/A^e	Y/Y^f	λ°/A^g
Abramovitz[h] 1869–1953	0.25		3.10		0.75				1.53				1.68	3.51	
Solow[i] 1909–49	0.35		1.77		0.65				1.09				1.49	2.90	
Kendrick[j] 1889–1953	0.28		2.71		0.72				1.72				1.60	3.60	
Denison[k] 1909–29	0.234		3.16		0.689	1.58	4.34	0.38	2.29 / 0.56	0.10	0.01	0.28	0.28	2.82	0.92
1929–57	0.225		1.88		0.73	1.31	−0.73	0.50	2.16 / 0.93	0.15	−0.01	0.34	0.39	2.93	1.70
Denison[l] 1950–62	0.144	3.74		3.00	0.78	1.14	4.21		1.42 / 0.62		−0.13	0.30	1.07	3.32	1.56
Kuznets[m] 1889–1929	0.33		3.76		0.67				1.74				1.24	3.70	
1929–57	0.23		1.01		0.77				0.53					2.30	2.95
1950–62	0.214		3.88		0.787				0.80					1.87	3.36

Jorgenson and Griliches[n]

Study / Period	w_K	$\dot{K}/K$	w_L	$\dot{L}/L$	$\dot{A}/A$	$\dot{Y}/Y$	$\dot{A}^*/A^*$	Additional values
Jorgenson and Griliches[n] 1950–62	0.414	4.09	0.585	1.22	1.03	3.47	1.76	3.14[o], 0.70[p], 0.25[q], 0.63[o], 0.75[p]
Kendrick[r] 1948–66	0.247	3.48	0.754	1.31	2.30	4.10		4.16[q]
Denison[s] 1929–76	0.151	2.17	0.81	1.27	0.89	2.98	1.49	2.48, 1.07, −0.63, 0.31, 0.65, −0.13, 0.27
Denison[t] 1929–82	0.154	2.44	0.804	1.33	0.76	2.92	1.33	2.52, 1.17, −0.64, 0.29, 0.65, −0.13, 0.26
Jorgenson, Gollop, and Fraumeni[u] 1948–79	0.399	2.98	0.601	1.73	0.81	3.42	2.37	1.04, 1.12, 0.61

a. w_K = share of capital

b. $\dot{K}/K$ = rate of growth of capital

c. w_L = share of labor

d. $\dot{L}/L$ = rate of growth of labor

e. $\dot{A}/A$ = estimated rate of technical progress

f. $\dot{Y}/Y$ = rate of growth of output

g. $\dot{A}^*/A^*$ = estimated rate of technical progress in the absence of quality adjustments.

h. Abramovitz (1956): K series: capital stock—includes land, structure, producers' durable equipment, and net foreign assets. L series: man-hours (labor force less estimated unemployment multiplied by standard hours). Y series: real net national product. Assumptions: constant-returns-to-scale, profit-maximization, and neutrality of technical progress.

i. Solow (1957): K series: Goldsmith's (1956) capital stock (excludes government, agriculture, and consumer durables), adjusted for percentage of labor force unemployed. L series: man-hours. Y series: real private non-farm GNP. Assumptions: constant-returns-to-scale, profit-maximization, and neutrality of technical progress.

j. Kendrick (1961): K series: real net capital stock (by asset type) multiplied by the rate of compensation = real capital services. L series: man-hours worked weighted by average hourly earnings in industry group. Y series: real net product (Kuznets' concept—national security version, private and public consumption outlay and net investment plus national security expenditures). Assumptions: constant-returns-to-scale, profit-maximization, and neutrality of technical progress.

k. Denison (1962): K series: privately owned capital. L series: Office of Business Economics (OBE) estimates of persons engaged in production, adjusted for work hours (quantity and quality effects), education, experience and age and sex composition changes. Y series: real national income. Assumptions: degree of returns to scale = 1.1, cost-minimization, and neutrality of technical progress.

l. Denison (1967): K series: nonresidential structure and equipment, inventories, dwellings, and international assets. (Denison shows that the total contribution of capital amounts to 25% of the growth in real national income. This figure includes contributions from dwellings of 7.5%, international assets of 1.5%, nonresidential structures and equipment of 12.9%, and inventories of 3.0%. The contributions reported here include only those from nonresidential structures and equipment and inventories, whereas the figures in table 4-2 represent Denison's estimate of the entire contribution of capital. Denison attributes 0.25% of the 3.32% growth in national income to growth in dwellings and 0.05% to growth in international assets.) L series: employment, adjusted for hours worked, education, and age and sex composition changes. Y series: real national income. Assumptions: degree of returns to scale = 1.1, cost-minimization, and neutrality of technical progress.

growth accounting exercises by selected authors for the United States over different periods. All of the authors included in the survey except E. F. Denison assume constant returns to scale and profit maximization.[20] Denison assumes that if all inputs increase by 1 percent, output increases by 1.1 percent, so that modest increasing returns to scale and cost minimization result. All of the authors assume, at least implicitly, neutrality of technical progress.

The results of the various growth accounting exercises, even over identical or similar periods, differ significantly. For example, the estimates of technical progress for the prewar period range between 0.3 and 1.7 percent per annum and those for the postwar period range between 0.8 and 2.3 percent per annum. The lowest estimated rate of technical progress is the 0.3 percent by Denison for the period 1909–29. The highest estimated rate of technical progress is the 2.3 percent by S. S. Kuznets for the period 1929–57 and by J. W. Kendrick for the period 1948–66.[21]

One potential source of the differences is the concept of aggregate real output used by the various authors: Both Abramovitz and Kendrick use

20. Denison (1962, 1967, 1979, 1985).
21. Kuznets (1971); Kendrick (1973).

Table 4-1 notes continued

m. Kuznets (1971): *1889–1929 and 1929–1957 estimates derived from Lithwick 1967.* K series: structure and equipment in 1929 prices. L series: man-hours. Y series: real GNP (Department of Commerce versions). Assumptions: constant-returns-to-scale, profit-maximization, and neutrality of technical progress. *1950–1962 estimates derived from Denison 1967. The factor shares reported here are inferred from Kuznets' estimates of technical progress, Denison's raise of growth of real output and input as well as his estimate of the relative factor shares.* K series: Denison's (1967) capital stock. L series: Denison's (1967) man-hours, excluding education adjustment. Y series: real national incomes. Assumptions: constant-returns-to-scale, profit-maximization, and neutrality of technical progress.

n. Jorgenson and Griliches (1967, 1972): K series: stock—consumers' durables, nonresidential structure, producers' durables, residential structure, nonform inventories, form inventories, and land; utilization (of business equipment and structures)—approximated by the utilization of power-driven equipment in manufacturing. L series: persons engaged, adjusted for changes in labor force composition by educational attainment and for effective hours per prerson. Y series: real gross private domestic product. Assumptions: constant-returns-to-scale, profit-maximization, and neutrality of technical progress.

o. Stock.

p. Quality.

q. Utility.

r. Kendrick (1973). K series: real capital stock weighted by sectoral capital compensation. L series: employment multiplied by hours worked, weighted by industry average rate of compensation. Y series: real net national product, adjusted for government productivity improvement. Assumptions: constant-returns-to-scale, profit-maximization, and neutrality of technical progress.

s. Denison (1979): K series: Inventories, weighted average of an index of gross and net stock of nonresidential structure and equipment, dwellings, and international assets (Denison attributes 0.17% of the 2.98% growth in national income to growth in dwellings and 0.02% to growth in international assets). L series: employment adjusted for hours, frequency, and quality effects), education, and age and sex composition changes. Y series: real national income. Assumptions: degree of returns to scale—1.1, cost-minimization, and neutrality of technical progress.

t. Denison (1985): K series: inventories, weighted average of an index of gross and net stock of nonresidential structures and equipment, dwellings, and international assets (Denison attributes 0.20% of the 2.02% growth in national income to growth in dwellings and 0.06% to growth in international assets). L series: employment adjusted for hours (quantity and quality effects), education, and age and sex composition changes. Y series: real national income. Assumptions: degree of returns to scale—1.1, cost-minimization, and neutrality of technical progress.

u. Jorgenson, Gollop, and Fraumeni (1957): K series: capital stock adjusted for changes in the composition of aggregate capital stock by asset class and legal form of organization. L series: hours worked adjusted for changes in composition of hours worked by age, sex, education, employment class, and occupation. Y series: aggregate real value-added. Assumption: constant-returns-to-scale, profit-maximization, and neutrality of technical progress.

real net national product; R. M. Solow uses real private nonfarm gross national product; both Denison and Kuznets for the period 1950–62 use real (net) national income; D. W. Jorgenson and Griliches use real gross private domestic product; Kuznets for the periods 1889–1929 and 1929–57 uses real gross national product; and Jorgenson, M. Gollop, and B. M. Fraumeni use real aggregate value added (approximately equivalent to real gross domestic product).[22] Thus the rates of growth of aggregate real output as used by the various authors may not be the same over identical or similar periods.

Moreover, because of the variations in the concept of aggregate real output, the measured share of labor also varies across studies—from approximately 0.6 if real output is measured gross of depreciation to approximately 0.8 if real output is measured net of depreciation. Under the maintained hypothesis of competitive profit maximization with respect to labor, the measured share of labor may be used as an estimate of the production elasticity of labor. Thus aggregate real output may be estimated to grow by between 0.6 and 0.8 percent for every 1 percent growth in labor input, other things being equal. Under the maintained hypothesis of constant returns to scale, the production elasticities sum to unity. Hence, the production elasticity of capital may be estimated as one minus the share of labor—between 0.2 and 0.4. For every 1 percent growth in the capital input, aggregate real output may be expected to grow by between 0.2 and 0.4 percent, other things being equal.[23]

These growth accounting exercises may be further distinguished by whether they incorporate quality adjustments in the measurement of the quantities of inputs. Abramovitz, Solow, Kendrick, and Kuznets do not make any quality adjustments to the quantities of measured inputs.[24] Their estimates of technical progress range from 1.24 to 2.30 percent per annum with an average of 1.78 percent. Jorgenson and Griliches; Denison; and Jorgenson, Gollop, and Fraumeni all make quality adjustments to the quantities of labor. Jorgenson and Griliches as well as Jorgenson,

22. Abramovitz (1956); Kendrick (1961, 1973); Solow (1957); Denison (1962, 1967, 1979, 1985); Kuznets (1971); Jorgenson and Griliches (1967, 1972); Jorgenson, Gollop and Fraumeni (1987).

23. Denison (1962, 1967, 1979, 1985) assumes that the degree of returns to scale is 1.1. Thus, the sum of his estimates of production elasticities of capital, labor, and land should sum to 1.1. Under the assumption of cost minimization, the production elasticities may be estimated as the respective cost shares multiplied by 1.1. Thus, the production elasticities of capital, labor, and land in Denison (1985) may be estimated as 0.17, 0.88, and 0.05, respectively.

24. Abramovitz (1956); Solow (1957); Kendrick (1961, 1973); Kuznets (1971).

Gollop, and Fraumeni make quality adjustments to the quantities of capital, too.[25] Their estimates of technical progress range from 0.28 to 1.07 percent per annum, with an average of 0.78 percent. Studies without adjustments for the changes in the quality of inputs, which include most of the early ones, have yielded much higher estimates of technical progress. On average, the estimated rate of technical progress in these studies is more than 100 percent higher than that found in studies with quality adjustments.

For the purpose of comparison, for those studies with quality adjustments, approximate implied estimates also were computed of the rate of technical progress if quality adjustments were not made. In every case, significantly higher estimates were obtained of the rate of technical progress, as expected. The average estimate is 1.59 percent per annum, comparable to that obtained for the studies with no quality adjustments. This is consistent with the view that changes in the qualities of the inputs may be considered as one of the forms in which technical progress is materialized. Quality improvements in capital goods may have resulted from R&D, and hence the finding that quality adjustment may be large may point to a larger contribution of R&D capital to growth than presumed in (all or part of) the residual.

In table 4-2, the estimated (implied) contributions are presented of the three sources of growth—capital, labor, and the residual (which is usually identified as technical progress)—for the various studies, with and without quality adjustments. On average, the proportion of economic growth attributed to the residual, with quality adjustment of the inputs, is 25 percent. This proportion increases to 52 percent in the absence of quality adjustments of the inputs. In other words, when changes in the quality of inputs are attributed to the residual or technical progress, its estimated contribution is on average doubled.

The proportion of economic growth explained by growth in the capital input is, on average, 28 percent with quality adjustment and 22 percent without quality adjustment. Quality adjustments thus increase the proportion of economic growth attributable to capital.

Whether quality adjustments to the inputs should be considered part of technical progress is debatable. To the extent that the origin of the quality improvements is unknown or unpredictable, a good case can be made that they belong in the residual just as much as other forms of

---

25. Jorgenson and Griliches (1967, 1972); Denison (1962, 1967, 1979, 1985); Jorgenson, Gollop, and Fraumeni (1987).

technical progress. Ultimately it depends on the objective of the particular growth-accounting exercise whether quality adjustments should be considered as part of technical progress or the residual.

Angus Maddison also performs a growth accounting for various periods and subperiods of 1913–84 for the G-5 countries plus the Netherlands.[26] He uses the growth accounting framework of Denison and others but examines a longer list of supplementary factors to explain the sources of gross domestic product (GDP) growth. He does not estimate these effects econometrically, and hence his estimates are subject to many of the problems raised above (for example, assuming neutrality of technical progress). However, he does include quality adjustments for labor and capital, small scale economies, and a catch-up effect to the lead country. For Japan, his residual is under 10 percent; for the United States, 22 percent for 1950–73 and zero for 1973–84.

In a series of studies, the contributions of capital, labor, and technology to economic growth in the G-7 countries in the postwar period have been recounted. A primary purpose of these studies is to explore the implications of relaxing the strong maintained hypotheses usually employed in econometric estimation of production functions and in growth accounting exercises: constant returns to scale, neutrality of technical progress, and profit maximization. These assumptions simplify the decomposition of growth into the contributions of various sources, but they can be enormously misleading if incorrect. They can be tested econometrically and can be rejected statistically. This leads to different findings on the contribution of capital, labor, and the residual to economic growth than in the traditional studies, as well as interesting observations on human capital and R&D.[27]

In table 4-3, the rates of growth of real output, measured as real gross domestic product, and the inputs—capital, human capital, labor, and R&D capital—for the G-7 countries are presented. Capital is measured as the cumulative total nonresidential fixed investment, in constant prices, by the perpetual inventory method. Labor is measured in hours, unadjusted for quality. R&D capital is measured as cumulative R&D expenditures, again in constant prices. Japan has the highest rate of growth of output as well as the highest rates of growth of capital and R&D capital in the sample period. By contrast, the United Kingdom has the lowest rate of growth of output and R&D capital. The United States,

26. Maddison (1987).
27. See Boskin and Lau (1990, 1991, 1992, 1995).

Table 4-2. *Contributions of Different Sources of Growth, by Authors and Time Periods*[a]

Percent

Authors and time periods	Time period with quality adjustment					Time period without quality adjustment				
	Capital	Technical progress	Subtotal	Labor	Scale	Capital	Technical progress	Subtotal	Labor	Scale
Abramovitz (1956) 1869–1953	n.a.	n.a.	n.a.	n.a.	n.a.	22	48	70	33	n.a.
Solow (1957) 1909–49	n.a.	n.a.	n.a.	n.a.	n.a.	21	51	72	24	n.a.
Kendrick (1961) 1889–1953	n.a.	n.a.	n.a.	n.a.	n.a.	21	44	65	34	n.a.
Denison (1962) 1909–29	26	10	36	54	10	26	33	59	32	10
1929–57	15	20	35	54	12	15	58	73	16	12
Denison (1967) 1950–62	25	32	57	34	9	25	47	72	19	9
Kuznets (1971) 1889–1929	n.a.	n.a.	n.a.	n.a.	n.a.	34	34	68	32	n.a.

Study										
1929–57	n.a.	n.a.	n.a.	n.a.	n.a.	8	78	86	14	n.a.
1950–62	n.a.	n.a.	n.a.	n.a.	n.a.	25	56	81	19	n.a.
Jorgenson and Griliches (1972 reply) 1950–62	49	30	79	21	n.a.	40	51	91	8	n.a.
Kendrick (1973) 1948–66	n.a.	n.a.	n.a.	n.a.	n.a.	21	56	77	24	n.a.
Denison (1979) 1929–76	15	30	45	46	9	15	50	65	26	9
Denison (1985) 1929–82	19	26	45	46	9	19	46	65	26	9
Jorgenson, Gollop, and Fraumeni (1987) 1948–79	47	24	71	30	n.a.	12	69	81	20	n.a.

n.a. Not available.

a. Some authors assumed the capital and labor shares are fixed; others, variable. In studies in which the shares are variable, the average over the sample period was used in the calculations. As a result, the contributions from the different sources may not sum.

Table 4-3. *Rates of Growth of Output and Inputs*

Percent per year

Country	Output	Capital	Human capital	Labor	R&D capital
Canada	3.90	5.20	0.90	1.80	5.70
France	3.40	6.20	1.30	−0.27	4.90
West Germany	2.90	4.90	1.00	−0.60	5.60
Italy	3.10	3.90	1.40	−0.20	5.70
Japan	5.80	8.20	0.80	0.60	8.60
United Kingdom	2.20	4.40	0.90	−0.30	2.10
United States	2.90	3.20	0.70	−1.80	4.50

Note: Canada, France, Italy, and Japan, 1964–90; West Germany and United Kingdom, 1965–90; United States, 1961–90.

with the second lowest rate of growth of output, has the lowest rate of growth of capital and human capital among the G-7 countries.[28] In all these studies, even with the inclusion of human capital, the residual remains large.

The Contribution of R&D to Economic Growth

The residual, captured by the time trend in our approach, is essentially a summation of the effects of all of the omitted factors that affect output. R&D capital is an important omitted factor. Supporting that claim, for example, J. J. Kim and L. J. Lau found that no technical progress was evident in the East Asian newly industrialized countries (NICs) of Hong Kong, Singapore, South Korea, and Taiwan, despite their very rapid rates of growth, in contrast to the finding for the G-5 or the G-7 countries.[29] The East Asian NICs have until recently not invested much in R&D. Many other productivity-enhancing expenditures are not considered capital investments, such as software development and organizational innovation, which have not been taken into account. The residual also could capture negative factors such as depletion of natural resources or increasing regulation.

What happens if R&D capital is introduced as a fourth input in the aggregate production function? The answers, using the metaproduction function approach, are in table 4-4. The introduction of R&D capital at the beginning of the period, unlagged, reduces the contribution of the

28. The level of human capital in the United States was already the highest among the G-7 countries at the beginning of the sample period.

29. Kim and Lau (1994).

Table 4-4. *Sources of Economic Growth*

Percent

Country	Capital	Labor	Human capital	R&D capital	Technical progress
Canada	20.5	23.1	2.8	10.0	43.6
France	42.5	− 4.1	4.8	11.6	45.2
West Germany	40.2	− 10.3	4.6	15.5	49.9
Italy	27.2	− 1.9	5.8	15.8	53.0
Japan	43.8	2.2	2.1	14.2	37.7
United Kingdom	49.8	− 5.2	4.9	8.3	42.1
United States	32.3	18.4	2.4	9.9	36.9

residual. However, it also reduces the contribution attributable to human capital, indicating possible complementarity between human capital and R&D capital. Even then, technical progress remains the most important source of growth, except in Japan and the United Kingdom.

In table 4-5, the effect of introducing R&D capital into the aggregate production function is compared with the estimates of the contribution of technical progress or the residual. A substantial reduction clearly occurs in the contribution of the residual. However, the contribution of the residual remains large. In other words, while the proportion of the contribution of the residual attributable to R&D capital is significant, it is relatively small.

Why is this the case? First, the residual embodies not only the effect of R&D but also other omitted factors and investments that do not typically enter into the capital account, such as learning-by-doing, software development, organizational innovations, market development, and other maintenance-type expenditures. Second, included in the production

Table 4-5. *The Contribution of the Residual and of R&D Capital to Economic Growth, 1961–90*[a]

Percent

Country	With capital, human capital, and labor	With capital, human capital, labor, and R&D capital
Canada	50.1	43.6
France	54.5	45.2
West Germany	64.8	49.9
Italy	69.7	53.0
Japan	53.3	37.7
United Kingdom	46.7	42.1
United States	50.7	36.9

a. Canada, France, Italy, and Japan, 1964–90; West Germany and United Kingdom, 1965–90; United States, 1961–90.

function is the R&D capital stock at the beginning of the period; to the extent that time lags exist between the expenditure on R&D and the realization of the benefits on a commercial scale, the effect of the current period R&D capital stock on the output of the current period should not be large. Thus, if lags are not introduced, the effect of R&D capital will be understated. Third, one cannot rule out the possibility that R&D capital is complementary to other forms of capital and therefore its effect on output may not be fully captured econometrically. However, this appears unlikely insofar as the contributions of neither physical capital nor human capital appear excessively large.

Three more factors may influence the identifiability of the effect of R&D at the macroeconomic level. First, in the aggregate, the social returns as well as private returns are fully reflected. The spillover effects, if any, are captured in the output of the aggregate economy. Thus, in principle, a higher rate of return to R&D is expected at the macroeconomic than at the microeconomic level. Second, many of the benefits of R&D are in the future, and to the extent the R&D capital continues to expand, the increase in output in the current period does not fully reflect the total benefits of R&D. It may be necessary to look at such data as the stock market valuation of firms to obtain a measure of the potential future contribution of intangible capital. Third, at the microeconomic level, however, successful R&D activities, especially in the private sector, tend to create market power, which in turn may lead to static inefficiencies in the economy, partially offsetting the beneficial effects of R&D for the aggregate economy. The race for these monopoly profits may increase R&D and innovation in the aggregate, as successive waves of intense competition for the technological advantages that produce the (temporary) monopoly profits generate more R&D, larger net aggregate innovation, productivity growth, and overall increases in the standard of living.

Conclusion

We share the view, expressed most thoughtfully in Griliches' American Economic Association's presidential address, that R&D is an important source of economic growth, but that the size of its effect econometrically is modest.[30] Its effect is likely somewhat larger than the numbers reported

30. Griliches (1994).

above (which are more or less consistent with earlier and other types of studies on the subject) for a number of reasons.

First, while numerous important case studies, historical analyses, and industry-level statistical analysis combine with appealing analytical conjectures to suggest that R&D has been an important contributor to growth from these points of view, and while in principle it should show up in the aggregate data, various problems prevent it from doing so. For example, much informal R&D is not measured in the attempts to get at R&D investment rates and R&D capital stocks. Much of the way that aggregate data are constructed might decrease the estimated effect of R&D on economic growth. For example, much R&D is sold to the government, a large sector in which output is measured by input cost, assuming no productivity growth. To the extent that R&D contributes to productivity growth, it will not be measured in the government sector. Only to the extent that R&D is similarly commercialized, and shows up as spillovers in increased output in some private sector activity in which output is measured directly or at least in a manner that allows for productivity growth, will this effect show up. Griliches discusses other important data problems in pinning down the contribution of R&D to productivity growth.[31]

A second issue relates to the extent to which measured productivity growth occurs because of quality change in inputs, especially tangible capital, that may have been heavily, or at least partly, the result of R&D. Attributing the growth to quality improvements begs the question. Looking at the partial derivative of productivity growth with respect to R&D, holding everything constant, to the extent that everything is measured properly and specified correctly, the direct effect of R&D in the production function studies could be established or its contribution estimated. But to know the total effect of an increase in R&D investment on output growth, its indirect effects via increases in the other sources of growth, including quality improvements in the capital stock (and perhaps, also, human capital), also must be taken into account.

Our analysis of the sources of U.S. economic growth suggests that serious attempts to include measures of quality change sharply reduce the residual. To the extent that the residual is identified as a measure of technical progress, and the even bolder additional step is made that much of the technical progress results from formal and informal R&D, a much

31. Griliches (1994).

smaller residual suggests there is less potential impact of R&D on productivity growth included in the now smaller residual.

Our findings in particular show the modest effects of R&D on the growth of output for the G-7 countries in the postwar period.

Typical estimates for the contribution of the growth of the R&D capital stock on output growth range from about 10 to 15 percent. However, large residuals also are found. But this suggests that not all of the residual is attributable to R&D as conventionally measured. The modest direct effect is not a large percentage of the residual, which suggests that while some additional, perhaps informal, R&D is one of the remaining explanations for the residual, there is probably far more in it than just R&D.

Our estimates of the direct effect of R&D may also be an underestimate. Because the economic benefits of R&D accrue well after the R&D investment is made, and because the R&D stock, while it depreciates, can be dipped into for subsequent commercial use, future economic benefits result. Only if the economy is in a steady state relative to all its variables, including R&D investment, would the accruing economic growth benefits to R&D from a ramping up of investment be matched by the retirement of benefits from previous R&D. Also, the lag structure, not the same as the ramp-up effect is still being explored and may lead to a small increase in the directly estimated effects of R&D.

Perhaps the most consistent result from all our studies, those estimating the effect of R&D, as well as those that excluded R&D, is that complementarity exists between R&D and human capital, human capital and tangible capital, and technology and tangible capital. These complementarity effects are substantial, given strong capital augmenting (in narrow or broad measures of capital, the latter inclusive of human capital or R&D capital) technical change. The higher the level of the capital stock, the larger is the effect of technical change, and vice versa. This suggests that the direct effect of the impact of the R&D as conventionally measured may also be an underestimate, as it interacts with human capital, tangible capital, and other aspects of technical change.

To take a simple example, the benefits of successful R&D in improved microprocessors to the economy depend upon, among other things, the amount of tangible capital that can benefit from better and faster microprocessors, the human capital necessary for people to be able to use the computers, and other forms of technology, such as advanced software, that can better utilize the capabilities of the better microprocessors. Ironically, a historical case is used to demonstrate the aggregate econometric results, thus coming full circle. R&D is important to economic growth, but just how important is a question economists are not yet fully able to answer.

Comment by Charles L. Schultze

These papers are so deep and rich that I can only dip in here and there. Let me start with the Richard R. Nelson and Paul M. Romer paper. I labor under the tremendous disadvantage that I essentially agree with Nelson and Romer's analysis and their policy conclusions. Their distinctions among hardware, software, and wetware are analytically very useful, and I endorse intellectual property rights as alternative ways of supporting knowledge generation. That discussion combines subtlety and realism with rigor. I further agree with them on the futility of the argument that government research support should be allocated solely on the basis of technological opportunities and should not be results- or target-oriented. In practice in the United States, federal support for research and development (R&D) has always been heavily targeted to meet particular demands. And now with defense R&D falling, some attention should be given to targeting research toward the industrial needs of the country. Most people do not realize just how targeted R&D in the United States has been and the extent to which it has been concentrated in just a few areas. In 1992, 89 percent of total federal R&D funds went to four sectors—defense, space, health, and energy. Even within basic research, 65 percent of federal R&D went to those four sectors. If defense is excluded, 72 percent of civilian research and development in the United States went to space, health, and energy and the remainder to everything else. The same proportions hold for basic civilian research. In the fiscal 1992 budget, 20 percent of the entire civilian R&D support was allocated to the space shuttle and the space station.

The U.S. federal support for civilian research and development outside of space, health, and energy is far lower than in other major countries. Classification problems exist, so the numbers are only broadly in the ballpark, but in Japan, France, the United Kingdom, and Germany on average 69 percent of government civilian R&D support is allocated to other areas such as health and energy, compared with 28 percent in the United States. Room clearly exists for reallocating funds away from current targets to other targets. And in turn, no good reason can be found for why those other targets should not include the broad economic or industrial needs of the country. Again, I agree with this main point of Nelson and Romer.

Now, however, I would like to suggest principles to observe in a national policy that explicitly targets governmental R&D support to the industrial needs of the country. First, do not specify the target as increasing American competitiveness. *Competitiveness* is a virtually meaningless, if widely used, word. It can—and has been—used to justify virtually anything. The purpose of international trade is to enjoy mutual benefits, not win a contest. As Paul Krugman has said so nicely, the purpose of trade is to import; unfortunately, exporting also is necessary because suppliers are so crass as to demand payment for the goods they sell.

Opportunities to add to the national income by exploiting other countries or cornering the market in high-tech industries are minimal. According to Krugman, if the United States made a major, and unopposed, effort to do just that, 1 percent might conceivably be added to the national income. The danger is that by specifying competitiveness as a national objective in allocating R&D support, attention will focus too much on manufacturing and too much on exports.

The goal ought to be to stimulate research and development that is likely to improve productivity in the American economy, whether in the 20 percent that is manufacturing or the 80 percent that is not; whether in the 85 percent that goes to produce domestically used goods or the 15 percent that goes to exports.

A similar trap to avoid is excessive concentration on high-tech industries. The U.S. Department of Commerce has identified ten such industries that are involved in organic chemicals, computers and electronics complex, and aerospace. Laura Tyson identifies essentially the same ones.

These high-tech industries allegedly provide external benefits to the economy, substantially more than other industries. They stimulate productivity elsewhere in the economy in a way that is not reflected in their own earnings. Innovation in an industry can lead to large gains elsewhere in the economy that are not captured in the private returns of the innovating industry. Edwin Mansfield has done a classic research project using seventeen case studies of innovations in which the social returns exceeded the private returns. The innovation with the highest excess of social return—overprivatization—was a new type of industrial trend; the second highest was a household cleaning device; the third highest, a new stain remover; and the fourth highest, a new construction material. The fifth highest is a high-tech example—computer-driven machinery innovation. The point is that R&D and innovation can pay large social benefits in the most prosaic of industries. Making better bricks can be terribly important. Therefore, a broad industrial distribution of targets is desired for that component of R&D geared to raising national productivity.

Subsidizing joint research projects should be done selectively and skeptically. The central nature of the problem addressed by R&D is reducing uncertainty. And early in the game many alternative paths ought to be explored. A national policy that rewards the combining of independent research efforts may do harm by prematurely closing off alternative paths. There should be a rebuttable presumption against subsidizing large R&D joint projects in which many firms combine their efforts. I am not suggesting that joint research projects should not be supported. The initial presumption ought to be not to, but that presumption can be rebutted with strong evidence. Joint projects are guilty until proven innocent. I suggest a paper by Linda Cohen and Roger Knoll, which analyzes the whole question.[1]

Another general principle needs to be observed in allocating R&D to economic targets. The U.S. Congress is constitutionally incapable of making choices among competing projects and cutting off funds for failed projects on the basis of cold-blooded, hard-headed economic criteria. As too often seen by political sponsors, the purpose of a project is primarily to generate jobs and income, not to produce industrial innovation. Therefore, specific project choice should be insulated as much as feasible from the political process. Within broad politically determined target areas, rely on peer review as much as possible: Choose individual projects. And the National Institutes of Health, with all its faults, is not a bad example to try to imitate.

Essentially, all of this is consistent with the Romer and Nelson view. National R&D policy should be flexible enough to realize that only a science push, a nontargeted approach, is at play. But the targeting process should be set up carefully, both substantively and politically. Otherwise, money will be wasted and harm will result.

The Michael J. Boskin and Lawrence J. Lau paper encompasses a large body of empirical work that the two authors have had under way for some time and have published in several installments. That body of work represents a major contribution to the professional literature on the sources of economic growth, though I have not had a full chance to absorb it. Let me try to pick and choose in a few places, and then I will go a little further than they have in one of their major findings.

1. Cohen and Knoll (1992).

The Boskin and Lau contribution is to construct a metaproduction function. Essentially they use data from not one country but a whole group of countries, and they combine time-series and cross-section data to get econometric estimates of the pace of technological progress. In concept, at least, this allows the data to determine several critical aspects about the production function that otherwise have to be assumed in advance in the traditional growth accounting.

The paper encourages the interested reader to continually ask, "Why don't we try this test—why don't we try that test? Here's an alternative use of the metaproduction function." In other words, the approach is a rich source of further work that could have many payoffs. I am going to raise a few questions about the work, but I do want to stress that I think that it makes an important contribution.

First, in terms of displaying results, more useful to the reader would be the contributions of various sources to the growth in output per worker instead of the growth of total workers. Two countries may have had similar elasticities of response of output to various kinds of inputs. But if one happened to have a much faster growth in the labor force, then labor would be a larger contributor to growth. When Boskin and Lau calculate the proportionate contribution to total output growth, a misleading interpretation results. Displaying contributions to the growth of output per worker would remove the effects of differential rates of labor for growth.

Second, I question the potential importance of scale economics, when dealing with large economies, once land is included in the production function. In a world that is increasingly globalized, limitations on firm size mean less and less. Does the size of Japan—roughly half that of the United States—have a major impact on the output of workers? I find it hard to conceptualize where the scale effects can come from, once a country has developed a very large economy.

Federal R&D spending in the United States, which accounts for a substantial fraction of all R&D spending, is concentrated in a few areas on the basis of criteria that have little to do with economic payoff. As Boskin and Lau recognized, in the three major areas of concentration—defense, health, and space—no independent measures of output or faulty measures exist. So I have grave doubts about what any macroeconomic analysis relying on these data can reveal about the potential contribution of R&D to economic growth.

As an aside, I have to say that this would also be true of the traditional growth accounting, as well as in the Boskin and Lau approach. But the miscalculation of the federal R&D is a major limitation. One might experiment with taking some of the federal R&D out of the R&D input.

Next, if the pace of technological progress changes substantially during a period of analysis, as it most probably did in the early 1970s in all of the seven countries that Boskin and Lau studied, the use of a single time trend to capture the residual generates an average set of results that may not be reflective of any one period and may produce input elasticities that do not capture the underlying process either at the beginning or the end. But most important, the question of convergence arises. If new technology is embodied in new capital goods, the contribution to growth investment will differ substantially, depending on whether a country is at the technological frontier, as the United States has been in the postwar period, or is in a catch-up phase, well below the frontier, as was the case in Europe and Japan in the first three decades of the postwar era. In the catch-up economy, a large backlog exists of available unexploited technological opportunities. A big rise in the investment share of gross domestic product can generate a long-lasting increase in the growth rate without exhausting that backlog and running into diminishing returns. In an economy at the frontier, however, higher investments today simply use up the much smaller backlog of technology much more quickly, and the contribution of investment to growth will be much shorter lived and less important. A cross-section involving economies at different catch-up stages may give some results with respect to factor elasticities that are inapplicable to an economy at the frontier.

Finally, I want to combine the Boskin and Lau conclusion that technological advances and human capital interact to strengthen each other with several other facts and draw a policy conclusion. Technology and human capital interact in a way that is characterized by diminishing returns. If human capital development stands still, the payoff from each unit of additional scientific, technological, and managerial improvement will decline. Strengthening this view is that technological advance not only consists of large technological breakthroughs but also is an accumulation, firm by firm and bit by bit, of translating and adapting the new science and technology and making a stream of small changes in the design of products and processes on the shop floor and in offices, all of which ultimately leads to improvements in quality and efficiency. The faster the technological change, the greater the adaptation required by the labor force.

Next, over the past twenty years, the distribution of wages in the United States has gotten steadily more unequal, with absolute losses for the bottom 30 to 40 percent of the work force. The demand for workers with higher education, skills, and competence rose relative to the demand for workers with lower qualifications. The weight, even if not the unanimity, of the economic research on this question suggests that this tilt in demand principally resulted from changing technology. At the same time, the supply of skills and competencies did not keep up with the demands of changing technology, and a surplus of low-skilled workers was created.

In the relatively flexible U.S. labor market, this surplus caused a reduction in wages at the bottom end of the work force. Unable to fill their skill requirements, firms down-skilled their production processes and hired the surplus workers at lower productivity and lower wage jobs. The same problem affected Europe. But European wages are rigid, and their safety net is more generous, so displaced workers are not forced to take lower wage jobs. The result is not lower wages but high unemployment. The failure of education and training to keep up with the demand for higher individual skills and competence, especially in the bottom half of the labor force, may now be acting as a bottleneck that reduces the economy's ability to translate technological advances into improved productivity.

If this is correct, it raises the question, in the context of very scarce federal budget dollars, as to whether efforts to improve education, training, and school-to-work transitions at the lower end of the labor force ought not to be given higher priority than additional dollars for R&D targeted to industrial purposes.

References

Abramovitz, M. 1956. "Resource and Output Trends in the United States since 1870." *American Economic Review* 46: 5–23.

Arrow, K. J. 1962. "The Economic Implications of Learning by Doing." *Review of Economic Studies* 2: 155–73.

Boskin, M. J., and L. J. Lau. 1990. "Postwar Economic Growth of the Group-of-Five Countries: A New Analysis." Technical Paper 217. Stanford, Calif.: Stanford University, Center for Economic Policy Research.

———. 1991. "Capital Formation and Economic Growth." In *Technology and Economics: A Volume Commemorating Ralph Landau's Service to the National Academy of Engineering*, 47–56. Washington: National Academy Press.

———. 1992. "Capital, Technology, and Economic Growth." In *Technology and the Wealth of Nations,* edited by N. Rosenberg, R. Landau, and D. Mowery, 17–55. Stanford University Press.

———. 1995. "A New View of Postwar G-7 Growth."

Denison, E. F. 1962. "United States Economic Growth." *Journal of Business* 35: 109–21.

———. 1967. *Why Growth Rates Differ: Postwar Experience in Nine Western Countries.* Brookings.

———. 1979. *Accounting for Slower Economic Growth: The United States in the 1970s.* Brookings.

———. 1985. *Trends in American Economic Growth, 1929–82.* Brookings.

Cohen, Linda, and Roger Knoll. 1992. "Research and Development." In *Setting Domestic Priorities,* edited by Henry Aaron and Charles Schultze, 223–66. Brookings.

Griliches, Zvi. 1988. "Productivity Puzzles and R&D: Another Non-explanation." *Journal of Economic Perspectives* 2 (4): 9–21.

———. 1994. "Productivity, R&D, and the Data Constraint." *American Economic Review* 84: 1–23.

Hall, Bronwyn H., and Robert E. Hall. 1993. "The Value and Performance of U.S. Corporations." *Brookings Papers on Economic Activity* (1): 1–50.

Jorgenson, D. W., and Z. Griliches. 1967. "The Explanation of Productivity Change." *Review of Economic Studies* 34: 249–83.

———. 1972. "Issues in Growth Accounting: A Reply to Edward F. Denison." *Survey of Current Business* 52 (5). Part II: 65–94.

Jorgenson, D. W., F. M. Gollop, and B. M. Fraumeni. 1987. *Productivity and U.S. Economic Growth.* Harvard University Press.

Kendrick, J. W. 1961. *Productivity Trends in the United States.* Princeton University Press.

———. 1973. *Postwar Productivity Trends in the United States, 1948–69.* Columbia University Press.

Kim, J. I., and L. J. Lau. 1994. "The Sources of Economic Growth of East Asian Newly Industrialized Countries." *Journal of the Japanese and International Economies* 8: 235–71.

Kuznets, S. S. 1971. *Economic Growth of Nations.* Harvard University Press.

Lau, L. J., and P. A. Yotopoulos. 1989. "The Meta-Production Function Approach to Technological Change in World Agriculture." *Journal of Development Economics* 31: 241–69.

Maddison, Angus. 1987. "Growth and Slowdown in Advanced Capitalist Economies: Techniques of Quantitative Assessment." *Journal of Economic Literature* 25: 649–98.

Romer, Paul M. 1990. "Endogenous Technological Change." *Journal of Political Economy* 98 (5): S71–S102.

Solow, R. M. 1957. "Technical Change and the Aggregate Production Function." *Review of Economics and Statistics* 39: 312–20.

CHAPTER 5

Contributions of New Technology to the Economy

Edwin Mansfield

T HE BASIC IDEAS underlying the techniques used by economists to measure the social benefits from technological innovations are by no means new; they stem from Alfred Marshall's work of a century ago on consumer surplus. Nor are they difficult to understand. Consider a new product (used by firms) that can shift the supply curve of the industry using the new product. How far downward this supply curve will shift depends on the pricing policy of the innovator. Assume that the innovator decides to set a price for its new product that yields a profit to the innovator equal to r dollars per unit of output of the industry using the innovation (for example, r dollars per appliance for a new type of metal used by the appliance industry). Also, assume that the industry using the innovation is competitive, that its demand curve is as shown in figure 5-1, and that its supply curve is horizontal in the relevant range. In particular, suppose that, before the advent of the innovation, this supply curve was S_1 in figure 5-1, and the price charged by the industry using the innovation was P_1. After the advent of the innovation, this supply curve is S_2, and the price is P_2.

The social benefits from the innovation can be measured by the sum of the two shaded areas in figure 5-1. The upper shaded area is the consumer surplus resulting from the lower price (P_2 instead of P_1) stemming from using the innovation. Also, a resource saving is registered, and a corresponding gain in output elsewhere in the economy, because the resource costs of producing the good using the innovation—including the resource costs of producing the innovation—are less than P_2Q_2. Instead, they are P_2Q_2 minus the profits of the innovator from the innovation, the latter being merely a transfer from the makers of the good using the innovation to the innovator. Thus, in addition to consumer

Figure 5-1. *Social Benefit from a New Product*

Price or cost per unit of output of
industry using the innovation

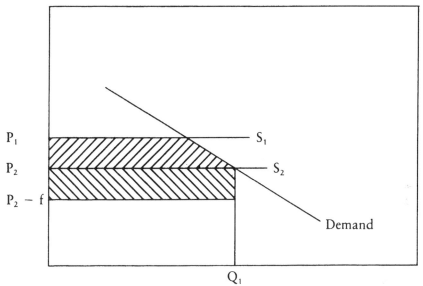

Output of industry using the innovation

surplus arising from the price cut, a resource saving is present amounting to the innovator's profits.

In many cases, two adjustments must be made in this estimate corresponding to the lower shaded area in figure 5-1. First, if the innovation replaces another product, the resource saving does not equal the innovator's profits (from the innovation), but these profits less those that would have been made (by the innovator or other firms) if the innovation had not taken place and the displaced product had been employed instead. This is the proper measure of the resource saving. Second, if other firms imitate the innovator and begin selling the innovation to the industry that employs it, their profits from the sale of the innovation must be added to those of the innovator to get a complete measure of the extent of the resource saving caused by the innovation.

The social benefits also can be measured from new products used by individuals instead of firms and from new processes. But given that the principles involved are much the same as those described above, the measurement procedures will not be presented here.[1]

1. A description is provided in Mansfield and others (1977a).

Social Rates of Return

A social rate of return is the interest rate received by society as a whole from an investment. To economists, the social rate of return from investments in new technology is important, because it measures the payoff to society from these investments. A high social rate of return indicates that society's resources are being employed effectively and that more resources should be devoted to such investments, if the rate of return stays high. In a series of papers, I have tried to describe the many difficulties in measuring and interpreting the social rate of return.[2] They are numerous and important, but until something better comes along, estimates of the social rate of return are likely to continue to be used.

Although earlier efforts to measure the social rates of return from investments had been made in agriculture, the first attempt to measure the social rate of return from investments in industrial innovations was published in 1977. The innovations that were included in the study were found in a variety of industries, including primary metals, machine tools, industrial controls, construction, drilling, paper, thread, heating equipment, electronics, chemicals, and household cleaners. They occurred in firms of different sizes. Most were of average or routine importance, not major breakthroughs. While the sample could not be viewed as randomly chosen, it was not obviously biased toward very profitable innovations (socially or privately) or relatively unprofitable ones. The findings indicated that the median social rate of return from the investment in these innovations was 56 percent, a very high figure (see table 5-1).[3]

To extend this sample and replicate the analysis, the National Science Foundation commissioned two studies, one by Robert R. Nathan Associates and one by Foster Associates. Their results, like those in table 5-1, indicate that the median social rate of return tends to be very high. Based on its sample of twenty innovations, Nathan Associates found the median social rate of return to be 70 percent. Foster Associates, based on its sample of twenty innovations, found the median social rate of return to be 99 percent.[4]

More recently, Manuel Trajtenberg estimated that the social rate of return to research and development (R&D) in the field of computed tomography (CT) scanners in medical technology was about 270 per-

2. Most recently, Mansfield (1991b).
3. Mansfield and others (1977a).
4. Foster Associates (1978); Robert Nathan Associates (1978).

Table 5-1. *Social and Private Rates of Return from Investment in Seventeen Innovations*

Percent

Innovation	Rate of return	
	Social	Private
Primary metals innovation	17	18
Machine tool innovation	83	35
Component for control system	29	7
Construction material	96	9
Drilling material	54	16
Drafting innovation	92	47
Paper innovation	82	42
Thread innovation	307	27
Door-control innovation	27	37
New electronic device	a	a
Chemical product innovation	71	9
Chemical process innovation	32	25
Chemical process innovation	13	4
Major chemical process innovation	56	31
Household cleaning device	209	214
Stain remover	116	4
Dishwashing liquid	45	46
Median	56	25

Source: Mansfield and others (1977a).
a. Negative.

cent.[5] As he is careful to point out, interpreting the gains as social depends on the motives underlying the behavior of hospitals when choosing medical technologies. Also, like hybrid corn, which Zvi Griliches studied, the rate of return would be expected to be high because the innovation was known in advance to be a gusher, not a dry hole.[6] But bearing these things in mind, Trajtenberg's results are consistent with the proposition that the social rate of return from investments in new technology tends to be high.[7]

A well-known, and hard-to-explain, slowdown has occurred in the rate of productivity increase, in the United States and abroad. While

5. Trajtenberg (1990).
6. Griliches (1958).
7. The findings in this section all pertain to the average rate of return, which may differ substantially from the marginal rate of return—the rate of return from an additional dollar spent. Attempts have been made to estimate marginal rates of return by the use of econometric estimation of production functions. In general, these estimates suggest that the marginal rates of return are very high. For a summary of these findings and their limitations, see Mansfield (1980 and 1991b).

some have blamed this on diminishing returns to science and technology, no proof exists that this is the case, or that a substantial drop has been evident in the social rate of return from investments in new technology.

Social versus Private Rates of Return

While the social returns from innovative activities seem to be very high, the private returns—the returns to the innovating firm—are not necessarily high. Whereas the median social rate of return was 56 percent, the median private rate of return was only 25 percent (see table 5-1).[8] Also, rich and detailed data were obtained regarding the return from the innovative activities (from 1960 to 1972) of one of the nation's biggest firms. For each year, this firm made a careful inventory of the technological innovations stemming from its R&D and related activities, and it made detailed estimates of the effect of each innovation on its profit stream. The average private rate of return from this firm's total investment in innovative activities was about 19 percent, which was much less than the social rate of return.[9]

To understand why the private rate of return from innovative activity is so much lower than the social rate of return, it is important to recognize that the innovator often has difficulty appropriating the returns from the innovation. Many benefits accrue to imitators, who often quickly obtain information concerning the detailed nature and operation of the new products and processes. According to a study of one hundred U.S. firms, this information is in the hands of at least some of their rivals within about a year, on the average, after a new product is developed (see table 5-2).[10] For processes, this information leaks out more slowly, but generally in less than about fifteen months. The major exception is chemical processes, which often can be kept secret for a number of years.

Technical information of this sort spreads in many ways. In some industries, personnel move from one firm to another, informal communications networks operate among engineers and scientists working at various firms, and professional meetings serve as information exchanges. In other industries, input suppliers and customers are major channels (because they transmit a great deal of relevant information), patent ap-

---

8. Mansfield and others (1977a).
9. Mansfield and others (1977b).
10. Mansfield (1985).

Table 5-2. *Percentage Distribution of U.S. Firms, by Average Number of Months (after Development), before the Nature and Operation of an Innovation Are Reported to be Known to the Firms' Rivals*

Industry	New products					New processes				
	<6	6-12	12-18	18+	Total	<6	6-12	12-18	18+	Total
Chemicals	18	36	9	36	100	0	0	10	90	100
Pharmaceuticals	57	14	29	0	100	0	33	0	67	100
Petroleum	22	33	22	22	100	10	50	10	30	100
Primary metals	40	20	0	40	100	40	40	0	20	100
Electrical equipment	38	50	12	0	100	14	14	57	14	100
Machinery	31	31	31	8	100	10	20	30	40	100
Transportation equipment	25	50	0	25	100	0	67	0	33	100
Instruments	50	38	12	0	100	33	33	33	0	100
Stone, clay, and glass	40	60	0	0	100	0	20	20	60	100
Other	31	15	15	38	100	27	0	36	36	100
Average	35	35	13	17		13	28	20	39	

Source: Mansfield (1985).

plications are scrutinized intensely, and reverse engineering is carried out. In still other industries, such information is not a closely held secret, partly because secrecy is regarded as futile in any event. Thus the intelligence-gathering process varies substantially from industry to industry (and from case to case).

The rapid diffusion of information does not mean that imitation will occur equally fast. Considerable time often is required to invent around patents (if they exist), to develop prototypes, to adapt or build plant and equipment, and to engage in the manufacturing and marketing startup activities needed to introduce an imitative product or process. But the basic information concerning the nature and operation of the innovation, even if not sufficient to allow the immediate introduction of an imitative product or process, is of great importance to the innovator's rivals. And if information is prone to find its way into the hands of rivals within a year or two, obvious and significant implications can be drawn concerning the incentives for innovation.

The extent to which the social benefits from R&D are appropriable depends on how much competition the potential innovator faces and on the kind of research or development activity in question. The more competition there is, and the more basic the R&D activity, the less appropriable the benefits are likely to be.

Patents and Lead Time

The patent system increases the extent to which the social benefits from innovations can be appropriated by their inventors. However, patents are of much more importance in a few industries such as pharmaceuticals and agricultural chemicals than in others. Frequently, imitators can invent around patents without too much difficulty. For example, one investigation found that, within four years of their introduction, about 60 percent of the patented successful innovations studied were imitated.[11]

According to R. Levin and others, firms generally regard lead time, moving quickly down the learning curve, and sales or service efforts as more effective than patents to protect the competitive advantage of new or improved products and processes.[12] The most effective way for a firm to appropriate a substantial share of the profits from its R&D investment is to exploit the head start it has over its rivals. It must commercialize its innovation rapidly and produce and market its innovation to impede the entry of potential rivals. Also, it must move quickly down the learning curve (which cuts its costs) and build strong links to distribution channels and customers.

Some firms can develop and commercialize new products and processes much more rapidly than others, thus increasing this vital lead time. Japanese firms in the chemical, rubber, machinery, instruments, metals, and electrical equipment industries have tended to develop and commercially introduce innovations more quickly than U.S. firms.[13] But the situation varies from industry to industry. In some industries, such as machinery, the time differential has been about 20 percent; in other industries, such as chemicals, it has been negligible (if it exists at all).

K. Clark, W. Chew, and T. Fujimoto have investigated the differences among European, Japanese, and U.S. auto producers in product development performance.[14] They, too, find that the Japanese have had an advantage in both lead time and engineering hours. According to their findings, the Japanese tend to overlap various development activities to a greater extent than U.S. or West European firms. Also, the relevant segments of a Japanese firm tend to exchange information more inten-

11. See Mansfield, Schwartz, and Wagner (1981); and for a study of the effects of intellectual property rights in an international context, see Mansfield (1995) and Lee and Mansfield (forthcoming).
12. Levin and others (1987).
13. Mansfield (1988b).
14. Clark, Chew, and Fujimoto (1987).

sively and informally than do comparable segments of a U.S. or West European firm. Further, Japanese suppliers play a much different role in the design process than most of their U.S. or West European counterparts, which may be important in explaining the Japanese advantage in lead time.

U.S. firms have responded to this challenge from Japan (and elsewhere). For example, according to a University of Michigan study, average development time in the U.S. auto industry fell from fifty-six months in 1991 to forty-eight months in 1993, and according to a Delphi survey, it may go down to forty-two months by 1998. The Japanese performance is also improving, but the gap between Japanese and U.S. performance seems to be narrowing.[15]

Industrial Innovation in Japan and the United States

In carrying out an innovation, Japanese firms have allocated their resources differently from the way U.S. firms do. One hundred Japanese and U.S. firms were examined to determine the proportion of the total cost of developing and introducing a new product that was incurred in each stage of the innovation process: (1) applied research, (2) preparation of project requirements and basic specifications, (3) prototype or pilot plant, (4) tooling and manufacturing equipment and facilities, (5) manufacturing startup, and (6) marketing startup (see table 5-3).[16] The proportion of total innovation cost devoted in Japan to tooling and

---

15. University of Michigan survey of three hundred executives, engineers, and auto analysts, summarized by J. Lippert of Knight-Ridder News Service, in *Philadelphia Inquirer*, March 26, 1994.

16. The first three stages include what is generally regarded as research and development. Applied research is aimed at a specific practical payoff. Preparation of product specifications consists of routine planning and scheduling, as well as cooperative work involving R&D, marketing, and other parts of the firm. The prototype or pilot plant is self-explanatory. Tooling and manufacturing equipment and facilities include the amounts spent before the first sale and delivery of the product on the preparation of detailed manufacturing drawings, toolings, and the design and construction of manufacturing facilities. Manufacturing startup includes the cost of training of workers, the debugging of production facilities, and the institution of acceptable quality control procedures, as well as the output produced before an acceptable quality level is reached. Marketing startup includes the cost of marketing studies undertaken to investigate the demand for the product, advertising campaigns before its introduction, establishment of a system of distribution, and training of the sales force in the operation or characteristics of the product. All innovation costs pertain only to the period before the first sale and delivery of the new product.

Table 5-3. *Percentage Distribution of Innovation Cost, One Hundred Japanese and U.S. Firms, 1985*

Stage of innovation process	Japan	United States
Applied research	14	18
Preparation of product specifications	7	8
Prototype or pilot plant	16	17
Tooling and manufacturing equipment and facilities	44	23
Manufacturing startup	10	17
Marketing startup	8	17

Source: Mansfield (1988a).
Note: Figures may not total 100 because of rounding.

manufacturing equipment and facilities was almost double that in the United States. This reflects Japan's emphasis on process engineering and efficient manufacturing facilities. On the other hand, the proportion of innovation cost going for marketing startup—that is, the expenses of pre-introduction marketing activities—in the United States was almost double that in Japan.

The quickness and efficiency of Japanese firms in utilizing and adapting external technology stems partly from their construction of relatively effective systems to monitor technological developments outside their own country. One hundred major U.S. firms in thirteen industries were asked to rank firms of five countries (France, West Germany, Japan, United Kingdom, and United States) in this regard (see table 5-4). The consistency with which the Japanese were ranked first is explained, at

Table 5-4. *Average Rank of Five Major Countries by the Perceived Effectiveness of Their Industry in Monitoring Technological Development outside Their Own Country*

Industry	France	Germany	Japan	U.K.	U.S.
Chemicals	3.8	3.0	1.5	4.4	2.2
Pharmaceuticals	4.2	3.1	1.4	4.2	2.0
Petroleum	3.6	2.2	1.2	4.6	3.0
Primary metals	4.2	2.7	1.0	4.6	2.5
Electrical equipment	3.8	2.9	1.0	4.0	3.3
Machinery	3.8	2.7	1.2	4.4	2.8
Transportation equipment	3.7	2.0	1.0	4.7	3.7
Instruments	4.1	2.6	1.4	3.7	3.3
Rubber	3.5	3.0	1.0	5.0	2.5
Stone, clay, and glass	4.0	3.6	1.0	3.8	2.6
Other	4.0	2.8	1.5	4.2	2.4
Average	3.9	2.8	1.2	4.3	2.8

Source: Mansfield (1990).
Note: Other includes fabricated metals, food, and paper.

Table 5-5. *Percentage of U.S. Firms Spending at Least as Large a Percentage of Sales on Monitoring Foreign Technological Developments as the Average Amount Spent by Their Foreign Rivals*

Industry	U.S. firms compared with firms in			
	France	*Germany*	*Japan*	*U.K.*
Chemicals	58	58	42	67
Pharmaceuticals	78	89	33	89
Petroleum	60	40	22	60
Primary metals	60	50	60	80
Electrical equipment	44	33	22	56
Machinery	20	20	10	40
Transportation equipment	67	33	33	67
Instruments	56	67	33	56
Rubber	0	0	0	100
Stone, clay, and glass	60	60	0	80
Other	58	67	67	75
Average	51	47	29	70

Source: Mansfield (1990).

least in part, by the systematic programs conducted by both Japanese firms and government agencies to learn about foreign technology. The effectiveness of the intelligence-gathering activities of a nation's firms depends in part on how much it spends on such activities. Less than 30 percent of the U.S. firms reported that they spent as much (as a percent of sales) on these activities as their Japanese rivals (see table 5-5).[17]

The Importance of Process Innovation

Innovations are often classified into two types: product innovations (new and improved products) and process innovations (new and improved processes). Some leading managers, engineers, and economists have warned that U.S. firms tend to neglect process innovation. While this claim is not easy to prove or disprove, Japanese firms have devoted about two-thirds of their R&D budgets to work on processes, whereas comparable U.S. firms have devoted only about one-third of their budgets to such work.[18] U.S. firms have allocated the bulk of their energy and resources to the development and introduction of new products, rather than new processes.

17. Mansfield (1990).
18. Mansfield (1988a).

Marie-Louise Caravatti, in a 1991 study for the International Trade Commission, found that about 80 percent of U.S. industrial R&D expenditure was aimed at product innovation, while only about 20 percent went for process innovation. In contrast, she cited a Japanese government survey of companies performing R&D during 1983–88, which found that only about 17 percent of Japanese firms said they were engaged in new product development, whereas 26 percent said they were trying to improve the technology of their own manufacturing processes, and 46 percent were trying to incorporate technology recently developed by others into their own production processes.[19]

Many major new products, such as the industrial robot, the videotape recorder, and the xerox copier, have originated in U.S. industrial laboratories. They have been of enormous significance to consumers and others, but frequently a large share of their profits have gone to non–U.S. producers, which have devised ways of making these products better and more cheaply than U.S. firms did. Many observers charge that this problem is caused in part by the relatively low status of manufacturing—relative to finance, marketing, research, and other functions of the firm—in some U.S. companies. The best engineering talent often has been attracted to research and design, not manufacturing. Positions in manufacturing have tended to have lower pay and less prospect for promotion than those in other areas. Relatively few chief executive officers of major U.S. corporations have come from manufacturing, unlike countries such as Japan and Germany, where manufacturing is much more highly esteemed.

Academic Research and Industrial Innovation

Industrial innovation has been mainly the province of engineers and scientists residing in firms. For example, nylon was based largely on Wallace Caruthers' work at Du Pont, and numerical control was pioneered by John Parsons, owner of a small firm that produced rotor blades. But this does not mean that academic research is unrelated to industrial innovation. On the contrary, based on data obtained from seventy-six firms in seven industries, about 10 percent of their new products and new processes could not have been developed (without substantial delay) in the absence of recent academic research; that is, academic

19. Caravatti (1992).

Table 5-6. *Percentage of New Products and Processes Based on Recent Academic Research, Seven U.S. Industries, 1975–85*

Industry	Percentage that could not have been developed (without substantial delay) in the absence of recent academic research		Percentage that were developed with very substantial aid from recent academic research[a]	
	Products	Processes	Products	Processes
Information processing	11	11	17	16
Electronics	6	3	3	4
Chemical	4	2	4	4
Instruments	16	2	5	1
Pharmaceuticals	27	29	17	8
Metals	13	12	9	9
Petroleum	1	1	1	1
Mean	11	9	8	6

Source: Mansfield (1991a).
a. See Mansfield (1991a, footnote 5).

research occurring within fifteen years of the commercialization of the innovation (see table 5-6).[20] The proportion of new products and processes based in this way on recent academic research was highest in the drug industry and lowest in the petroleum industry. Many innovations based on recent academic research were not invented at universities. Academic research often yields new theoretical and empirical findings and new types of instrumentation that are needed for the development of a new product or process, but it seldom results in the specific invention itself.

For new products and processes introduced in 1975–85 that, according to the innovator, could not have been developed (without substantial delay) in the absence of recent academic research, information was obtained concerning the mean time interval between the relevant academic research result and the first commercial introduction of the product or process. (If more than one such research result was needed for the development of the innovation, this time interval was measured from the year when the last of these results was obtained.) The mean time lag in these

20. Developing new products and processes without the findings of recent academic research is sometimes possible, but more costly and time consuming. Cases of this sort are designated in table 5-6 as ones in which development occurred with very substantial aid from recent academic research. Although firms technically could have developed them without the findings of recent academic research, it often appeared economically unwise to have attempted to do so. Thus, in a practical sense, many of these innovations could not have been developed (without substantial delay) in the absence of recent academic research. Substantial delay means a delay of a year or more.

industries was about seven years. (In interpreting this finding, note once again that these data pertain only to recent academic research.)[21]

In industries such as drugs, instruments, and information processing, the contribution of academic research to industrial innovation has been substantial. In seven industries, new products first commercialized in 1982–85 that could not have been developed (without substantial delay) in the absence of recent academic research accounted for about $24 billion of sales in 1985 alone (see table 5-6). And in these industries, new processes first commercialized in 1982–85 that could not have been developed (without substantial delay) in the absence of recent academic research resulted in about $7 billion in savings in 1985 alone. While these figures are rough, they indicate that innovation in these industries has been based to a substantial degree on recent academic research.

If the model in figure 5-1 is extended, rough estimates of the social rate of return from academic research can be made, based on these figures. The estimates are likely to be conservative for a variety of reasons, including the fact that the educational value of academic research is ignored. Nonetheless, the results indicate that the social rate of return exceeds 20 percent, even when generous assumptions are made concerning the contributions of industrial R&D, plant and equipment, and start-up activities.[22] Thus, even when viewed in a constricted light, the social payoff from academic research seems to have been high, although the limitations of such estimates should be stressed.

Sources of Academic Research Underlying Industrial Innovations

Seventy firms were approached to determine what kinds of academic research they believe were most important in producing innovations.

21. The average time lag will be much less for innovations based on recent academic research than for all innovations based on academic research, no matter when in the past the academic research occurred. Using data concerning the numbers of scientific papers published and the numbers of industrial scientists and engineers, Adams (1990) has estimated that an average lag of about twenty years occurs between the appearance of research in the scientific community and its effect on productivity in the form of knowledge absorbed by firms.

22. Even if one assumes that the social rate of return from industrial R&D, plant and equipment, and startup costs is more than 50 percent per year, an extremely high figure, the social rate of return from academic research is estimated to be in excess of 20 percent per year; see Mansfield (1992a).

Each firm was asked to cite about five academic researchers whose work in the 1970s and 1980s contributed most significantly to the firm's new products and processes in the 1980s. Eventually, usable data were obtained from sixty-six of the seventy firms in the sample. Because these firms account, on the average, for about a third of the R&D expenditures in these industries, the sample seems adequate. Taken as a whole, the sixty-six firms cited 321 academic researchers.

In most industries, the most frequently cited universities are world leaders in science and technology. For example, MIT, University of California at Berkeley, University of Illinois, Stanford University, and Carnegie-Mellon University are most often mentioned in electronics. But not all of the most frequently cited universities are world leaders in the relevant fields. Thus Washington University and the University of Utah, neither of which is among the top dozen departments of chemistry (according to ratings by the National Academy of Sciences), are both among the most frequently designated by the chemical firms. The bulk of the cited academic research occurred in departments closely related to the technology of the industry in question. In the electronics industry, more than 60 percent of the cited academic researchers were in electrical engineering or mechanical engineering departments. In the chemical industry, almost 70 percent were in chemistry or chemical engineering departments.

One factor that influences how frequently a particular university is cited is the quality of the university's faculty. Another factor is the scale of a university's R&D activities in the relevant area: A critical mass of researchers and equipment is often regarded as essential to achieve high productivity in particular aspects of academic research. Still another factor is the geographical proximity of a university to the firms in the sample. Because of the advantages of firms working with, and keeping abreast of developments at, local colleges and universities, schools located near many of the firms were mentioned relatively often.

Recognizing that the effects of these independent variables may differ from one industry to another, a statistical analysis was carried out separately in each of the five industries where the sample size is reasonably large. The results suggest that all three independent variables have the expected effects in practically all cases, and in about half of the cases they are statistically significant. However, considerable variation exists among universities in the extent to which the number of citations can be explained by these factors, which is not surprising both because of imperfections in the data and because these factors are not the only ones influencing the dependent variables.

Financial Support for Academic Research Underlying Industrial Innovations

Information was obtained from each cited academic researcher concerning the sources of his or her financial support.[23] Because the data pertain to all academic research carried out by each cited researcher during the 1970s and 1980s, they indicate the overall contours of a researcher's support, not just the support of a particular project. Because various parts of a researcher's portfolio of projects often are interrelated and the firms often cited more than one project by a researcher, this was the best way to proceed.

Practically all of the cited academic researchers had some government support for their research. In about two-thirds of the cases, it came, at least in part, from the National Science Foundation (NSF). The Department of Defense was also important, particularly in electronics, and the National Institutes of Health played a major role in supporting academic researchers cited by the health-related industries, especially pharmaceuticals. The Department of Energy and the National Aeronautics and Space Administration also provided substantial, but more limited, support. The federal government provided about two-thirds of the funding for the cited academic researchers (see table 5-7).

Industry, too, provided support; more than four-fifths of the cited academic researchers got research funds from industry. But industry generally supported a substantially smaller percentage of the total research budgets of the cited academic researchers than did government (22 percent versus 64 percent). As might be expected, firms were more important as sources of support for the cited academic researchers than for all academic researchers in the relevant fields. The percentage of R&D expenditures in colleges and universities financed by private sources was about 10 percentage points lower, on the average, than the percentage of the cited academic researchers' budgets financed by industry.

23. Correspondence was initiated with all of the cited academic researchers still alive in the United States. Eventually, after telephone and other follow-ups, data were obtained (wholly or in part) from more than 90 percent of them. Thus, the response rate is very high for a survey of this sort; see Mansfield (1995).

Table 5-7. *Federal Financial Support for Cited Academic Researchers and for All Academic R&D in Relevant Field*

Industry citing the academic researcher	Percentage of cited academic researchers whose research was funded (wholly or in part) by the federal government[1]	Mean percentage of research budgets of cited academic researchers funded by the federal government[a]	Percentage of academic R & D in relevant field funded by the federal government[b]
Electronics	95	69	75
Information processing	91	51	73
Pharmaceuticals	91	85	61
Chemicals	96	71	74
Petroleum	100	56	58
Metals	89	44	n.a.
Instruments	100	72	n.a.
Mean	95	64	68

Source: Mansfield (1995).

n.a. Not available.

a. The figures pertain to 1970–89. For those researchers who were involved in academic research during only part of this period, the figures pertain only to the part. For those who were involved in academic research during the entire period and whose pattern of support was significantly different during the 1980s than during the 1970s, the pattern of support during the 1970s was used because this was generally the period when the work occurred for which the academic researcher was cited.

b. For the electronics industry, the relevant field is electrical engineering; for information processing, computer science; for pharmaceuticals, life sciences; for chemicals, chemistry; and for petroleum, chemical engineering. The figures pertain to 1983 and come from National Science Foundation (1985).

In most cases, the government-funded work of the cited academic researchers preceded their industry-funded work, which often was aimed at deepening and extending their previous government-financed work. A considerable number of the cited academic researchers reported that their sources of financial support shifted substantially from the 1970s to the 1980s. In the latter decade, more of their funding came from industry, less from government.

Industrial Support of Academic Research

During recent years, industrial support of academic research has been growing rapidly (see table 5-8). Thus understanding the factors influencing the probability that a firm will support R&D at a particular university has become increasingly important. Holding other factors constant, this probability would be expected to be inversely related to the distance from the firm to the university. The less the distance, the easier and cheaper it is for academic and firm personnel to interact and work together on a face-to-face basis. Interview studies suggest that this factor may be im-

Table 5-8. *Industry-Funded R&D Expenditures at Universities and Colleges, 1972–91*

Millions of dollars

Fiscal year	Expenditures	Fiscal year	Expenditures
1991	1,216	1981	292
1990	1,134	1980	236
1989	998	1979	194
1988	872	1978	170
1987	790	1977	139
1986	700	1976	123
1985	560	1975	113
1984	475	1974	95
1983	389	1973	84
1982	337	1972	74

Source: National Science Foundation (1993).

portant, but some R&D executives now claim that progress in telecommunications has reduced its significance.

If distance is held constant, the chance that a firm will support R&D at a particular university would generally be expected to be directly related to the quality of the university's faculty in the relevant department, up to some point. But beyond this point, higher faculty quality may not be worth the additional costs it requires. Some kinds of R&D can be conducted about as well by a merely good scientist or engineer as by a leading light. Some of the most prestigious departments may impose conditions on industrial support that are far more onerous than those imposed by less prestigious departments at more modestly rated universities.

To test these and other hypotheses, detailed data were gotten from nine major firms that together account for about 15 percent of total R&D expenditures in the chemical, computer, petroleum, and pharmaceutical industries. Each firm was requested to choose at random two or three important types of academic R&D it supported during the 1980s and early 1990s and to estimate the probability that it would fund a project of this type at universities at various locations (less than 100 miles, 100–1,000 miles, and more than 1,000 miles away) and with various levels of faculty quality (good-to-distinguished, adequate-to-good, or marginal).[24] To estimate these probabilities, reference was made

24. A type of R&D means a category of R&D project such as a particular kind of polymer synthesis or catalysis research. In defining such a type, firms were asked to include reasonably similar projects that were of importance to the firm. The precise boundaries of

to the proportion of the firm's R&D support (in dollars) for projects of this type going to universities with these designated characteristics during this period. For all firms combined, data were gotten for twenty types of R&D, two of which are basic research (from the firm's vantage point and based on the NSF's definitions), the rest being applied R&D.[25]

Averaging over all types of R &D (but separating basic research from applied R&D), as expected, distance matters. Holding faculty quality constant, the chance that a firm would fund these types of R&D at a university less than 100 miles away is more than double the probability that it would fund them at a university located 100–1,000 miles away, and generally more than triple the probability that it would fund them at a university more than 1,000 miles away. But distance matters more for universities with only adequate-to-good or marginal faculties than for those with higher rated faculties. For the former universities, the chances of funding are very low unless they are within 100 miles of the firm.

Faculty quality is also important. If distance is held constant, the chance that a firm would fund R&D at a university with a good-to-distinguished department in the relevant field is higher than at a university with an adequate-to-good department, which in turn is higher than at a university with a marginal department. But the effects of faculty quality are much less for applied R&D than for basic research. About a 50-50 chance exists that these firms would fund applied R&D at a university where the relevant department is rated only adequate-to-good or marginal. This stems in part from a feeling that less prestigious departments are more likely to stick closely to the firms' problems and needs, rather than push the work in directions of more interest to the faculty. Further, the belief is that much of this applied R&D can be done satisfactorily at such departments.

The Role of Universities

A controversy is brewing over how far universities should go to promote the productivity and competitiveness of U.S. industry. The weakness of

---

each type were established by the firms because they were familiar with the nature of their projects and the interrelationships among them. National Academy of Sciences (1982).

25. Lee and Mansfield (1995).

U.S. industry has been much more in product and process development and in the commercialization of R&D than in research. Universities are not likely to play a central or direct role in the development of new products or processes. Most new products and processes that could not have been developed (without substantial delay) in the absence of academic research were not invented at colleges or universities; instead, academic research provided new theoretical and empirical findings and new types of instrumentation that were required for the development of the new product or process, but not the specific invention itself.

This seems unlikely to change. Successful product and process development demands an intimate knowledge of the details of particular markets and production techniques as well as the ability to recognize and weigh commercial and technical risks that can come only with first-hand experience. Universities do not have this expertise, and expecting them to get it is unrealistic. The essence of successful development is the ability to work effectively with production and marketing departments; the interface difficulties are well known and sometimes formidable. To attempt to include universities as major players in this process, particularly given the importance of timely decisionmaking, seems unrealistic.

As industry leaders have emphasized repeatedly, one of the principal roles of universities in the process of technological change is to provide well-trained students.[26] What is less frequently stressed is that this extends beyond science and technology. Some of the recent weaknesses of U.S. industry can be laid at the door of the nation's business schools, which tended to underemphasize technology in their curricula and which too frequently failed to forge strong and effective ties with schools of engineering.

While universities should continue to emphasize the training of students and the performance of basic research, there is no reason that they should not move toward closer alliances with industry. Nor is there any reason that universities with particularly close ties to certain industries and firms should not move further in this direction than others. Considerable advantages are found in exposing academicians to industrial problems. One of the oft-cited characteristics of U.S. universities is their diversity. No single formula is likely to apply to the wide variety of universities. Some will almost certainly go much further than others in redefining their missions as direct support of industrial research fostering the productivity and international competitiveness of U.S. industry. One

26. Rosenberg and Nelson (1994).

potential advantage is that a substantial amount can be learned about the costs and benefits involved, and the risks are likely to be small. A little experimentation may be a good thing.[27]

Exploration of Interfirm Differences in Extent of Linkage with Academic Research

This symposium is a sequel to one held on essentially the same topic more than twenty years ago.[28] In the paper that I gave at that earlier symposium, I listed a number of topics that, in my opinion, were in need of more investigation.[29] On rereading my earlier paper, I am heartened to report that a great many of these gaps have been filled, at least partially. I would like to suggest two topics that, from the vantage point of today, seem worthy of further study. This is only a small sample, but it is all I have time or space to include.

Within each of the industries in table 5-6, substantial differences exist among firms in the percentage of new products that could not have been developed (without substantial delay) in the absence of recent academic research. Consider, for example, the chemical industry. According to executives of the firms in the sample, this percentage was 15 or more in about one-fourth of the chemical firms and zero in about one-fourth of the chemical firms. Similarly, substantial differences are found among firms in each of these industries in the percentage of new processes that could not have been developed (without substantial delay) in the absence of recent academic research (see table 5-9). Firms with relatively high values of these percentages seem to have established closer and more effective links to academic research than firms in the same industry with relatively low values of these percentages. Their new products and processes seem to have been more interwoven with, and dependent on, academic research than those of their rivals with lower values of these percentages. However, the latter firms seem to have paid less attention to, or avoided dependence on, recent academic research.

Why was the percentage of new products and processes based on recent academic research so much higher in some firms than in others in the same industry? To what extent are differences among firms attribut-

27. Mansfield (forthcoming).
28. National Science Foundation (1972).
29. Mansfield (1972).

Table 5-9. *Lowest and Highest Quartiles of Firms' Percentages of Products and Processes Based on Recent Academic Research, Seven U.S. Industries, 1975–85*[a]

Industry	Products		Processes	
	Lowest quartile[b]	Highest quartile	Lowest quartile[a]	Highest quartile
Information processing	0	43	0	10
Electronics	0	45	0	10
Chemical	0	15	0	9
Instruments	2	50	0	9
Pharmaceuticals	0.25	60	0.25	75
Metals	2	17	0	25
Petroleum	c	c	c	c
Mean	1	38	0	22

Source: Author's calculations.
a. Products and processes that could not have been developed (without substantial delay) in the absence of recent academic research.
b. In industries where 25 percent or more of the firms report that the percentage is zero, the lowest quartile equals zero.
c. Too few petroleum firms are included in the sample to calculate quartiles.

able to size or R&D intensity? To some firms being more specialized (niche players) than others? To the location of some firms in close proximity to one or more major universities, while other firms are more isolated geographically? To the educational background of the firm's leading executives and top technical personnel? To the participation of some firms, but not others, in university-industry cooperative R&D programs of various kinds? And why did some firms, but not others, participate in such programs?

Studies aimed at answering these and other such questions should be of considerable interest to analysts and decisionmakers. A widespread feeling exists in the United States and elsewhere that the rate of technological change—and the pace of economic progress—depends on how effectively industry makes use of the findings of academic research, but little is known about the characteristics of those firms that are particularly effective in this regard.

Industrial Consulting by Academic Personnel

Besides contracts and grants to universities, an important avenue by which university expertise is mobilized to help transfer information and know-how to firms is industrial consulting by individual faculty members. Interviews with scores of industrial executives provide convincing evidence that the consulting activities of university personnel are extremely valuable. Moreover, most of the academic researchers cited by firms as having contributed significantly to their new products and processes had continuing consulting relationships with at least some of the firms supporting their academic research, and their students have taken jobs with at least some of these firms. In all industries other than drugs, more than half of the cited academic researchers reported that the problems they worked on frequently or predominantly stemmed from their industrial consulting. (The cited academic researchers' government-funded work frequently was based on ideas and problems they encountered in industrial consulting.)

Given the importance of consulting in the process of information transfer between universities and industry, little is known unfortunately about activities of this sort. Firms might be willing to provide information concerning the total amounts they have paid academic consultants for work related to their R&D activities. Also, data concerning the relative amounts spent on various kinds of consulting activities could be obtained. Besides helping to solve particular technical problems facing firms, academic consultants are sometimes enlisted to describe their findings or to brief company personnel on recent developments in a particular field or area of research. Also, they sometimes are useful to firms in evaluating students as potential employees and in bringing promising students to their attention. Without more detailed exploration of such consulting activities, analysts will miss some of the significant aspects of university-industry cooperative efforts.

Conclusions

Ten conclusions can be drawn from the studies I have reviewed.

First, practically all of the available evidence indicates that the economic payoff to society as a whole from investments in new technology has been very high, the average social rate of return being about 50 percent. No assurance exists that this high rate of return will be maintained as more investments are made, but thus far there is no convincing evidence that the marginal social rate of return has fallen substantially.

Second, the economic payoff to the innovator is frequently much lower than to society as a whole. Information concerning a new technology leaks out more quickly than is generally recognized. In other than a few areas such as pharmaceuticals and agricultural chemicals, the patent system frequently has only a limited effect on the rate of imitation. For investments in which the private rate of return from investments in new technology is low (even though the social rate of return is relatively high), a case may be made for public support.

Third, while research is of great importance, alone it is of limited economic significance. The contribution of research to a nation's economic performance depends on how well the nation's firms can utilize and commercialize research (internal and external) to bring about profitable new products and processes. As the case of Japan demonstrates, process engineering, tooling and manufacturing equipment and facilities, and quality control can be extremely important. While R&D is significant, what matters from an economic point of view is whether R&D can be integrated with marketing, production, and finance, effectively and in a timely fashion.

Fourth, investments in new technology tend to be risky. Forecasting the outcomes of fundamental research is notoriously difficult. But even if the relevant R&D project is not very risky, large commercial risks frequently must be taken. In other words, even if a firm can be reasonably sure of achieving a particular technological result, a sizable chance exists that this result will not be profitable to exploit. Analysis can take the innovator only so far, after which the new process or product must be subject to the test of the market, with the attendant risks.

Fifth, academic research has become a major underpinning for industrial innovation in many science-based industries. In information processing, pharmaceuticals, and instruments, among others, more than 10 percent of new products or processes in recent years could not have been developed (without substantial delay) in the absence of recent academic research. While the problem of estimating the social rate of return from academic research is extraordinarily difficult, the available data indicate that this rate of return is high, even if generous assumptions are made concerning the social rate of return from industrial R&D, plant and equipment, and startup activities.

Sixth, universities differ greatly in the extent of their contributions to industrial innovation. How frequently a university department is cited by firms as having carried out research that contributed importantly to their new products and processes depends on the quality of the department's faculty (as measured by the National Academy of Science ratings), the scale of the university's R&D activities in the relevant areas, and the extent to which firms in the industry are located nearby. However, a great deal of the interuniversity variation cannot be explained by these three factors; many other variables are also relevant.

Seventh, practically all academic researchers cited by firms for their contributions to the firms' innovations had some government support for their research. The National Science Foundation, the National Institutes of Health, and the Department of Defense were responsible for much of this support. The National Institutes of Health played a key role in supporting academic researchers cited by the health-related industries, particularly pharmaceuticals. The Department of Defense was particularly important to electronics, but it helped many other industries as well. The National Science Foundation's support seemed to go across the board.

Eighth, industry provided financial support to most of the cited academic researchers, but it was much smaller and arrived later than government support. An important factor that promoted the transfer of information and know-how between industry and the academic researchers was consulting. Most of the cited academic researchers reported that the problems that they worked on in their academic research frequently or predominantly developed out of their consulting for industry.

Ninth, while industry still provides a small proportion of the financial support for academic research, its share is growing rapidly. Based on data obtained from a small sample of firms with large R&D budgets, industry tends to fund research at nearby universities. Whereas some R&D executives claim that advances in telecommunications have reduced the significance of location, geography still is important, based on these data. Noteworthy from the viewpoint of regional development, the geographical location of academic research activity is likely to influence the extent and direction of technological change in the areas near the universities doing the work and the nature and growth of production and employment there.

Tenth, while university departments with highly rated faculties are more likely to receive industry support than those with lower rated faculties, the differences frequently are not large, and a substantial proportion of the academic research findings that industry regarded as most important in the 1980s came from modestly rated departments. Fortunately for society, a wide range of colleges and universities, not just the more prestigious ones, play an important role in fostering industrial innovations.

Despite the decades of work by economists and others that have gone on since the 1972 NSF symposium, a full, or even adequate, understanding of the role of science and technology in the economy is still far away. Unquestionably, advances have been made, but enormous gaps remain.[30]

References

Adams, J. 1990. "Fundamental Stocks of Knowledge and Productivity Growth." *Journal of Political Economy* 98: 673–702.

Caravatti, M. 1992. "Why the United States Must Do More Process R&D." *Research-Technology Management* (September–October).

Clark, K., W. Chew, and T. Fujimoto. 1987. "Product Development in the World Auto Industry." *Brookings Papers on Economic Activity.*

Foster Associates. 1978. *A Survey on Net Rates of Return on Innovation: Report to the National Science Foundation* (May).

Griliches, Z. 1958. "Research Costs and Social Returns: Hybrid Corn and Related Innovations." *Journal of Political Economy* 66 (October): 419–31.

Lee, J., and E. Mansfield. 1995. "Industrial Support of Academic Research." Paper presented at the annual meeting of the American Economic Association.

———. Forthcoming. "Intellectual Property Protection and U.S. Foreign Direct Investment." *Review of Economics and Statistics.*

Levin, R., and others. 1987. "Appropriating the Returns from Industrial Research and Development." *Brookings Papers on Economic Activity* 3: 783–831.

Mansfield, E. 1972. "Contribution of R&D to Economic Growth in the United States." *Science* 175 (February 4): 487–94.

———. 1980. "Basic Research and Productivity Increase in Manufacturing." *American Economic Review* 70 (December): 863–73.

———. 1985. "How Rapidly Does New Industrial Technology Leak Out?" *Journal of Industrial Economics* 34 (December): 217–23.

———. 1988a. "Industrial R&D in Japan and the United States: A Comparative Study." *American Economic Review* 78 (May): 223–28.

———. 1988b. "Industrial Innovation in Japan and the United States." *Science* (September 30).

———. 1990. "Comments on Productivity Growth in Japan and the United States." In *Productivity Growth in Japan and the United States,* edited by C. Hulten. Cambridge, Mass.: National Bureau of Economic Research.

———. 1991a. "Academic Research and Industrial Innovation." *Research Policy* 20: 1–12.

———. 1991b. "Estimates of the Social Returns from Research and Develop-

30. Bronwyn H. Hall's article in this volume covers the work of Zvi Griliches and his colleagues at the National Bureau of Economic Research. For a collection of recent papers on this topic by a wide variety of economists, see Mansfield and Mansfield (1993).

ment." In *Science and Technology Policy Yearbook,* edited by M. Meredith, S. Nelson, and A. Teich. Washington: American Association for the Advancement of Science.

———. 1992a. "Academic Research and Industrial Innovation: A Further Note." *Research Policy* 21: 295–96.

———. 1992b. "Appropriating the Returns from Investment in R&D Capital." In *European Industrial Restructuring in the 1990s,* edited by K. Cool, D. Neven, and I. Walter. London: Macmillan.

———. 1995. *Intellectual Property Protection, Direct Investment, and Technology Transfer: Germany, Japan, and U.S.* World Bank.

———. 1995. "Academic Research Underlying Industrial Innovations: Sources, Characteristics, and Financing." *Review of Economics and Statistics* (February): 55–65.

———. Forthcoming. "Links between University Research and Industrial Innovation." In *A Productive Tension: University-Industry Research Collaborations in the Era of Knowledge-Based Economic Development,* edited by P. David and E. Steinmueller. Stanford University Press.

Mansfield, E., M. Schwartz, and S. Wagner. 1981. "Imitation Costs and Patents: An Empirical Study." *Economic Journal* 91 (December): 907–18.

Mansfield, E., and E. Mansfield, eds. 1993. *The Economics of Technical Change.* Aldershot, England: Elgar.

Mansfield, E., and others. 1977a. "Social and Private Rates of Return from Industrial Innovations." *Quarterly Journal of Economics* 91 (May): 221–40.

———. 1977b. *The Production and Application of New Industrial Technology.* W. W. Norton.

National Science Foundation. 1972. *Research and Development and Economic Growth/Productivity.* Government Printing Office.

———. 1985. *Academic Science/Engineering: R&D Funds, 1983.* Government Printing Office.

———. 1993. *Academic Science and Engineering: R&D Expenditures, Fiscal Year 1991.* Government Printing Office.

Robert Nathan Associates. 1978. *New Rates of Return on Innovations: Report to the National Science Foundation* (July).

Rosenberg, Nathan, and Richard R. Nelson. 1994. "American Universities and Technical Advance in Industry." *Research Policy* 23: 323–48.

Trajtenberg, M. 1990. *Economic Analysis of Product Innovation: The Case of CT Scanners.* Harvard University Press.

CHAPTER 6

The Private and Social Returns to Research and Development

Bronwyn H. Hall

T HE PRINCIPAL argument for government intervention in industrial innovation has always been the potential gap between the private and social returns to innovative activity.[1] During the more than twenty years since the National Science Foundation Colloquium on R&D and Economic Growth/Productivity in 1971, a large amount of research effort has been expended both on measuring the extent of the gap and on evaluating efforts to close it via government policy. I will survey what has been learned from this research, focusing on the microeconomic evidence, and leave to others the task of integrating the evidence into a macroeconomic perspective. Even restricting my effort to firm or industry-level research, it remains a formidable task, and I will rely in some cases on research that went into some detail on particular topics.[2] In addition, I will confine my review to empirical evidence; theory is included only to the extent that it helps to frame the questions to be asked or to interpret the evidence.

The channels by which the benefits from innovative activity may spill over to agents other than those undertaking it are several. Although they have often been enumerated by other authors, repeating them is worthwhile.[3] First, firms in the same or related industries as an innovating firm may benefit through reverse engineering of products, the hiring away of scientists and engineers involved in innovation, or simply increased gen-

1. Nelson (1959); Arrow (1962).
2. See, for example, Mairesse and Sassenou (1991) and Mairesse and Mohnen (1995) on the contribution of R&D to productivity growth; Griliches (1992) and Mohnen (1994) on the measurement of spillovers and externalities; Terleckyj (1985) on the economic effects of federal R&D; and Cohen and Levin (1989) on market structure and innovation.
3. See, for example, Mansfield and others (1977, pp. 144–66) and Griliches (1992).

eral knowledge of the technology in question. The strength of these spillovers is likely to be a function of proximity, either in technology or geographic space. Second, to the extent that innovative firms are competitive (unable to behave as discriminating monopolists), firms and consumers that buy new products from an innovating industry may benefit by acquiring goods at prices lower than their willingness to pay for such goods. Third, research undertaken in the public or nonprofit sector that is freely disseminated will benefit innovating firms in that the cost of any particular innovation is reduced.

The role of these spillovers at the level of the individual firm has been studied to a greater or lesser extent during the past twenty years. This paper surveys the evidence on the flows from government and university-based research to individual firms, and from the firms to overall productivity growth. The focus is on the impacts of government spending in this area.[4]

Measurement Problems

Several important factors confound attempts to make precise measurements of the private and social returns to research and development (R&D) at the firm and industry level. None is a precisely new concern, but being reminded of them is useful. These factors are (1) the effect of price index (price deflator) measurement on the measurement of productivity growth, (2) the low variability of R&D spending in individual firms and the difficulties that creates for identifying the intertemporal aspects of knowledge production, and (3) the importance of R&D depreciation estimates for measuring rates of return.

In his article "Issues in Assessing the Contribution of R&D to Productivity Growth," written in 1979, Zvi Griliches outlined the difficulties of interpreting firm-level returns to R&D when price indices are poorly measured. The problems he describes are particularly severe with respect to new product innovation. If anything, these problems have worsened, as new and improved products have become an increasingly important part of the output of R&D.[5] Two implications can be drawn about the

4. For spillovers between and among firms, see the survey by Griliches (1992).

5. I have not been able to find statistical evidence on the trends in the relative proportions of industrial R&D devoted to new and improved products as opposed to new and improved processes, but casual observation and a glance at the major R&D–performing industries suggests that this proportion has risen over time. Mansfield (1988) presents

measurement of the contribution of R&D to productivity growth at the firm level. First, if the price indices do not adjust adequately for quality improvement in the output of the firm, measured output will grow too slowly and the contribution of R&D to its growth will be underestimated. Second, if the input price indices do not adjust adequately for quality improvement in the inputs to production, the measured input will grow too slowly and the contribution of R&D to productivity will be over-estimated.

To quote Griliches, "Conventional productivity measures reflect, therefore, the cost-reducing inventions made in the industry itself, the privately appropriated part of product innovations within the industry, and the social product of inventions in the input-producing industries which have not already been reflected in the price of purchased inputs."[6] The extent to which this statement is true depends on the kind of price deflation carried out before estimation is performed.

A second measurement difficulty that has confounded researcher after researcher as they explore lag structures in the impacts of research on output is that R&D spending is a smooth series at the firm level. That is, measuring and describing the lags between spending and productivity growth using econometric methods has proved almost impossible.[7] A related implication is that distinguishing between statements such as "R&D intensive firms tend to have higher productivity on average" and "Raising R&D investment increases the productivity of a firm" has proved extremely difficult.

Provided one is careful about the interpretation of results with different deflators, estimating the marginal revenue elasticity of R&D at the firm level with some precision is possible; except for its potential heterogeneity across firms, this measure is probably the easiest to obtain.[8] The reason is that this measure is a function of (relatively) well-measured quantities: real sales and R&D spending.[9] The difficulties arise when turning this measure into a measure of the returns to R&D.

evidence that this proportion is much higher in the United States than in Japan (two-thirds as opposed to one-third), which makes the issue particularly important in the U.S. context.

6. Griliches (p. 99, 1979).

7. See Hall, Griliches, and Hausman (1984) and Lach and Schankerman (1988) for evidence on the low within-firm variance of R&D.

8. The marginal revenue elasticity is the percentage increase in the sales of a firm in a particular year that can be attributed to a percentage increase in R&D spending. In principle, this should be the percentage increase in the present discounted value of sales over all future years that arise from the increase in R&D in one year.

9. In this case, failing to use a firm- or industry-specific deflator for output is the right

Consider a simple stylized model of the intertemporal effects of R&D on firm revenue:

$$(6\text{-}1) \qquad V(0) = \sum_{t=0}^{\infty} \beta^t \left[S(...,R_{-1},R_0,R_1,...,Rt,X_t) - R_t - Xt \right],$$

where R is R&D spending, X is spending on other (variable) inputs, β is a discount rate, and $V(0)$ denotes the present discounted value of this program of R&D spending. Note that the revenue function $S(\cdot)$ is written as a function of all past R&D inputs; no particular pattern has been imposed on their productivity. In this framework, the returns to R&D spending in any year are the partial derivative of firm value with respect to that year's R&D spending:

$$(6\text{-}2) \qquad \frac{\partial V}{\partial R_0} = \sum_{t=0}^{\infty} \beta_t \frac{\partial S(t)}{\partial R_0} - 1.$$

The marginal cost of a dollar of R&D spending is one dollar in the current year, and the marginal benefit is the present discounted value of the marginal contribution of that dollar to sales in all subsequent years. If the world were stationary (if the returns structure were stable over time) and R&D stopped contributing to revenue after several years, this form of the returns function could be estimated from a simple regression of sales on the past history of R&D.[10] Note that depreciation of R&D is implicit in specifying this relationship, but that it is estimated, not assumed.

However, most researchers go further than the equation above, partly because of the difficulties arising from the high correlation across years of R&D spending at the firm level. Two approaches are possible; both involve assuming the existence of some sort of R&D capital within the

---

thing to do. If one is interested in measuring the marginal revenue product to the firm (and ultimately, private returns), adjusting the output for the effects of lower prices would be a mistake, because the firm is not receiving those benefits. Alternatively, if the R&D is producing improved goods, enabling the firm to charge higher prices (that is, increasing product differentiation and, hence, the firm's market power), deflating by an increasing price index will remove some of the revenue gains that accrue to the firm from its R&D. Either way, if the concept to be measured is revenue elasticity (or the profit elasticity), it is incorrect to deflate output by anything more than a gross domestic product (GDP) or consumer price index (CPI) deflator.

10. This is because the coefficients of current sales on lags of R&D would be the same as the partial coefficients of sales in subsequent years on this year's R&D if the relationship were stationary.

firm, assuming a depreciation rate for this capital, and constructing a measure of the capital from a declining balance formula:

$$(6\text{-}3) \qquad\qquad K_t = (1 - \delta) K_{t-1} + R_t,$$

where K is the R&D capital and δ is its depreciation rate. Given such a capital measure, the revenue function $S(\cdot)$ is now rewritten as a function of R&D capital instead of the infinite stream of past R&D expenditures. The marginal revenue product of R&D in a given year becomes

$$(6\text{-}4) \qquad\qquad \frac{\partial S_t}{\partial R_0} = \frac{\partial S_t}{\partial K_t} \frac{\partial K_t}{\partial R_0} = \gamma \frac{S_t}{K_t} \frac{\partial K_t}{\partial R_0},$$

where γ is the elasticity of sales with respect to R&D capital. Substituting equation 6-4 into equation 6-2 results in

$$(6\text{-}5) \qquad \frac{\partial V}{\partial R_0} = \sum_{t=0}^{\infty} \beta^t \, \gamma \, \frac{S_t}{K_t} \frac{\partial K_t}{\partial R_0} - 1 = \sum_{t=0}^{\infty} \beta^t \, (1 - \delta)^t \, \gamma \frac{S_t}{K_t} - 1.$$

To go any further, depreciation rates, discount rates, and the sales-to-R&D capital ratio must be assumed to be roughly constant over time. Under these assumptions the net excess return to R&D spending is the following:

$$(6\text{-}6) \qquad \frac{\partial V}{\partial R_0} = \frac{\gamma}{r + \beta\delta} \frac{S}{K} - 1 = \frac{\rho}{r + \beta\delta} - 1,$$

where r is the discount rate $(1 - \beta)$ and ρ is simply dS/dK, the marginal revenue product of R&D capital. When ρ is equal to the discount rate plus the discounted depreciation rate, the net excess returns are zero, as expected.[11]

Equation 6-6 is used to obtain rates of return to R&D in two distinct ways in the literature: the first estimates the marginal revenue product r directly as dS/dK, and the second estimates the elasticity γ and multiplies it by the sales-to-capital ratio S/K to obtain an estimate of r. The first method has the advantage that it assumes equalization of gross rates of return (the increase in sales from increases in knowledge capital) across firms, which may be a more plausible assumption than the equalization of

---

11. The treatment abstracts from the effects of corporate tax system; taking account of the special tax treatment of R&D will change the cost of R&D capital but will not affect the basic point.

sales or output elasticities. However, more plausible might be to assume that the net returns are equalized. This is not the same thing unless the rental price of R&D capital faced by all firms is the same. Note the role of the depreciation rate δ in the denominator of equation 6-6.

The second method has the advantage of not being sensitive to the choice of depreciation rate until the final step, when γ is multiplied by S/K.[12] In the absence of separately measured depreciation rates for the output of R&D spending, making definitive statements about net rates of return to R&D spending is difficult, although measuring revenue elasticities fairly well is possible.

The three issues discussed here by no means exhaust the list of difficulties with the production or cost function approach to measuring the returns to R&D, but they are the most important that arise within that framework. Difficulties with the measurement of other inputs or worries about returns to scale and imperfect competition are secondary, at least for this purpose; the former because it has not had a major impact on the estimates (except via the previously mentioned deflation route), and the latter because they are easily accommodated within the framework.

Private Returns to R&D

Private returns to R&D at the firm level provide a good illustration of the effects of the measurement problems on conclusions reached about the contribution of R&D to economic growth. The productivity growth slowdown of the 1970s produced a wave of research exploring the contribution of R&D to productivity growth. Much of that research has been summarized in a series of survey papers.[13] Using data through the end of 1977, the consensus estimate of the R&D elasticity in these studies was about .10 to .15 in the cross-section, and somewhat less than that over time within a firm (equal to zero using one-year growth rates and

---

12. To the careful reader, this statement should not be obvious. γ has been defined as the elasticity of sales to R&D capital K, and in principle, the choice of depreciation rate in computing K should affect its estimate. However, numerous researchers have demonstrated that the logarithmic form of the production function is not sensitive to the choice of depreciation rate and that estimates of γ hardly change as δ is varied. (For example, see Griliches and Mairesse (1984) and Hall and Mairesse (1995).)

13. Mairesse and Sassenou (1991); Lichtenberg and Siegel (1991); Mohnen (1992).

as high as .09 using average growth rates). Results on the private rate of return to R&D were extremely variable.

Unfortunately, the surveys are often not explicit about the type of output deflation that was used; in principle, whether the results are interpreted as measurements of private returns depends crucially on deflation. In many cases, these studies were conducted using a single manufacturing sector deflator, so that the elasticity computed is a real sales elasticity and not an output elasticity. The implication thus is that substantial private returns accrued to being an R&D–intensive firm during the 1960s and 1970s.

Three studies of the contribution of R&D to productivity growth take the data through the end of the 1980s: Griliches in 1993 and Eric J. Bartelsman in 1990 using industry data and Bronwyn H. Hall and Jacques Mairesse in 1995 (building on work in Hall's 1993 paper) using firm data. Griliches as well as Hall and Mairesse demonstrate that conclusions about the magnitude of the R&D output elasticity rest on whether or not the output of the computing industry is deflated by the Commerce Department's new hedonic price deflator for computers.[14] Table 6-1 displays the results of firm-level and industry-level total factor productivity growth regressions, with and without Standard Industry Classification (SIC) 357 (office and computing equipment). In the industry-level regressions, the deletion of a single three-digit industry lowers the gross rate of return to R&D from 33 percent to 12 percent.[15] In the firm-level regression, the output elasticity falls by a factor of ten. The reason is that the price deflator for SIC 357 falls by 80 percent between 1981 and 1989, inducing a substantial measured increase in the output of this industry. Because the industry also has high R&D intensity and increasing R&D budgets, the output increase is explained by R&D.

Two conclusions can be drawn from the data in table 6-1 and the papers from which they are derived. First, the excess revenue elasticity

14. Griliches (1993); Bartelsman (1990b); Hall and Mairesse (1995); Hall (1993b). Bartelsman does not explicitly investigate this question, but his estimate of the R&D output elasticity is based on data that have been deflated by the new hedonic deflator, and the estimate is consistent both with Griliches (1993) and Hall and Mairesse (1995). (The within-firm estimate for company-funded R&D is .180 (.012) in table 5 of Bartelsman (1990b)).

15. Griliches (1993) also made an attempt to correct the inputs (semiconductors and components) to this industry for incorrect deflation and also to deflate pharmaceuticals by a properly constructed hedonic index. These corrections raise the R&D intensity coefficients in table 6-1 to .461 (.070) and .348 (.070), respectively, but they leave the gap resulting from deleting the computer industry essentially unchanged.

Table 6-1. The Returns to R&D

	Firm-level	Firm-level	Industry-level
Years	1981–89	1981–89	1978–89
Number of observations	7,616 or 8,110	5,967 or 6,520	142 or 143
R&D variable	R&D or growth in R stock	R&D-to-sales ratio	R&D-to-sales ratio
With computers	.108 (.006)	.269 (.058)	.330 (.073)
Without computers	.012 (.006)	.202 (.057)	.115 (.062)
With computers, no deflation	.027 (.006)	.231 (.055)	

Source: Mairesse and Hall (1994); Compustat data; Griliches (1993); National Bureau of Economic Research 4-digit database; and author's calculations based on Compustat data.
Note: The data are the estimated coefficient of R&D in a TFP regression, with and without firms in Standard Industry Classification (SIC) 357. In the first column, the R&D variable is the first difference of the log of R&D spending (which is approximately equal to the first difference of the log of R&D capital), and the coefficient measures of the elasticity of output with respect to R&D. In the other two columns, the R&D variable is the ratio of R&D to lagged sales, and the coefficient measures the gross return to R&D directly.

for R&D spending at the firm level appears to have declined toward zero during the 1980s. This by itself is not surprising; R&D is no longer a major source of sustainable rents. In his 1986 article reporting the results of an investigation into the private returns to R&D in the 1970s, Griliches said, "R&D as a major component of firm activity was undergoing a diffusion process in the 1950s and 1960s and may not have reached full equilibrium by the end of our period." By the end of the 1980s, at least a temporary equilibrium in the market value of this R&D had been reached.[16]

Second, the computer industry is an anomaly only in that it is one of the few industries for which a serious attempt has been made to adjust for quality change in the official price indices. To measure the output effects of R&D spending at the firm or industry level, the impact of new and improved products on prices in all industries, not just computing equipment, must be taken seriously. This is not a new point, but its importance has grown over the last ten to fifteen years.

Private Returns to Public R&D

Research and development spending covers a wide range of activities: basic laboratory research, or research aimed at the advancement of scientific knowledge, with or without commercial objectives; applied research directed toward practical applications; and research directed toward the development and production of specific new products and processes. The innovative activity itself encompasses even more activities, such as the identification of potential commercial opportunities, assessment of technical feasibility, marketing studies and research, the construction of new manufacturing facilities, and so forth. Nowhere is this more apparent than when considering the impact of federal R&D spending on private industry. More than half of the federal R&D budget goes to the Defense Department; of that 90 percent is spent on development, most of it in private industry (see table 6-2). The rest of the federal R&D budget is split roughly equally among health research, the National Aeronautics and Space Administration (NASA), energy research (including "big" science), and miscellaneous categories (in order of importance, the National Science Foundation (NSF), Departments of Agriculture, Interior, Commerce, and Transportation, the Environmental Protection

16. See Hall (1993a, 1993b) for evidence on this point.

Table 6-2. *Federal Funds for R&D, 1991*

Estimated, in millions of current dollars

Agency	Federal intramural	Industry (including FFDRCs)	Universities (including FFDRCs)	Nonprofits, other government	Total
Defense	8,988	25,640	1,693	596	36,917
Health	1,879	417	4,979	1,613	8,888
NASA	2,573	4,263	1,234	250	8,320
Energy	428	2,674	2,593	312	6,007
Other	2,528	580	2,346	522	5,976
Total	16,396	33,574	12,845	3,293	66,107

Source: National Science Foundation (1991, table 4-10).
FFDRC = Federally Funded Research and Development Centers.
NASA = National Aeronautics and Space Administration.

Agency, and so forth).[17] Thus looking for the impact of federal R&D is not likely to be a productive activity if only a single metric is used.

Consider the contrast between two examples of federally funded research projects: a Defense Department project to improve the technology of flat computer screens, mostly via grants to private industry, and Department of Energy funding for the construction of a new supercolliding accelerator (now canceled). In principle, the benefits from the first project are likely to flow primarily to firms in the industry (in addition to benefiting national defense) and possibly to consumers if there is price competition.[18] The benefits for society from the second project would be far more diffuse. Direct benefits arising from an increase in knowledge of the structure of matter are likely to take extremely long (decades or longer) to appear and to be exceedingly difficult to trace back to their source.[19] The more immediate impact comes as a by-product of the basic research activity. In their discussion of the supercollider project and its possible benefits, Paul A. David, David Mowery, and W. Edward Steinmueller identified three important by-products of basic research: (1) the education of scientists; (2) the creation of social networks through which

17. See National Science Board (1991, table 4-8).

18. Although contracts may be let to ensure market power for the firm(s) undertaking research (for example, allowing patent waivers so that the firms benefit from the patents they take out), this market power is likely to be greatly weakened, both by the fact that strong foreign competition exists already in this technology and by the rapid evolution that has been characteristic of technologies in this area.

19. See Rosenberg (1994) for some examples of the slow diffusion between fundamental scientific knowledge and important innovations that use that knowledge (for example, laser technology).

information diffuses rapidly before publication; and (3) the stimulus to technology and advances in instrumentation and techniques.[20] Although these effects may not have as long and variable a lag in their impact on growth, they, too, will be hard to measure except at fairly aggregate levels.

Thus two major research questions arise: The first asks specifically about the private returns to federal R&D performed within the firm (about half of all federal R&D is performed within industry; see table 6-1 for a breakdown). That is, does the R&D funded by the federal government act as a subsidy to the firm, or does it simply generate products that are demanded by the government without enhancing the firm's performance in other markets? The second and more difficult question concerns the spillovers to private firms from the part of federal R&D that is performed by governments, universities, and nonprofit research institutions (including FFRDCs, or Federally Funded Research and Development Centers). How large are these spillovers, and does their existence reduce the amount of R&D that the firms would otherwise undertake?

As part of the data collection effort that generates the aggregate statistics reported in the Science and Engineering Indicators, issued biennially, the National Science Foundation collects data on the R&D spending of a comprehensive sample of U.S. corporations and on the share of that spending that is funded by the federal government. Since 1972 microeconomic studies have repeatedly demonstrated that federally funded R&D generates a direct return of zero for the firms that do it, either at the firm or the industry level. Using data from U.S. firms, studies have been conducted by Griliches (firm-level data from 1957 to 1965), Griliches and Frank Lichtenberg (industry-level data from 1959 to 1976), Bartelsman (industry-level data from 1958 to 1986), and Lichtenberg and Donald

---

20. David, Mowery, and Steinmueller (1988, 1992). See also Rosenberg (1982) for a discussion of the role of basic research in developing and improving scientific instruments. In the specific case of elementary particle physics, Brooks (1985) identifies the technologies of massive data processing and analysis, high-precision surveying, mechanical design, cryogenics, high-power electric transmission, radio-frequency engineering, electronic engineering, control systems engineering, and large volume ultra-high vacuum design as those in which particle physics research has produced the impetus that led to advances. (On a personal note, I spent five years as a computer programmer in this field and can be considered an example of a spillover. At the time of my shift into econometrics programming in 1970, economists had just begun to use the large-scale datasets whose analysis had been familiar to elementary particle physicists for at least ten years and had not yet begun serious use of the nonlinear estimation methods that were routine in particle physics.)

Siegel (firm- and establishment-level data from 1972 to 1985).[21] Although superficially similar, the firm and industry studies sometimes differ significantly in the way in which federally funded R&D enters. In most of the firm-level studies and in the industry-level study of Bartelsman, the R&D in question is the R&D conducted by the firm but funded by the government.[22] Griliches, Lichtenberg and Siegel, and Bartelsman found zero or negative excess returns from this R&D.[23] Other researchers have found that the major impact of federal R&D spending at the firm level may be to increase the firm's own R&D spending.[24]

Griliches and Lichtenberg used the figures for R&D applied to particular product classes (27) that are collected by NSF; these product classes correspond approximately to an "industry of use" rather than "industry of origin" definition, so the experiment is fundamentally different from that being conducted at the firm level, where "industry of origin" numbers are being used.[25] Even so, they were able to find much evidence of a positive impact of federal R&D in an applied product field on the total factor productivity growth in the corresponding 2 +−digit industry. If anything, the contribution appeared to be negative in the more R&D−intensive sectors, and zero in others.[26] Taken at face value, these results suggest that the impact of federal R&D on cost reduction or productivity growth may be too diffuse to be captured even at the two-digit level.

The same result has been found using firm-level data in several other countries.[27] In addition, in a cross-country study of fifty-three countries, Lichtenberg found the contribution of government-funded R&D to be zero in a TFP regression that also contained privately funded R&D. The result is somewhat stronger in that it includes potential spillovers across

21. Griliches (1980); Griliches and Lichtenberg (1984); Bartelsman (1990a); Lichtenberg and Siegel (1991).

22. Although the discussion in Bartelsman suggests that the federal R&D data he uses is assigned to product fields (industry of use), the data available to him are only broken down into industries in which the R&D is performed. Zvi Griliches, private communication, 1995.

23. Griliches (1980); Lichtenberg and Siegel (1991); Bartelsman (1990a).

24. Mansfield (1984); Scott (1984); Terleckyj (1985); Lichtenberg (1985).

25. Griliches and Lichtenberg (1984).

26. Using a cost function approach, Mamuneas and Nadiri (1993) have conducted a similar exercise with data on twelve manufacturing industries between 1957 and 1988. They used a single aggregate federal R&D figure to construct public R&D capital as an infrastructure variable and included this variable in a set of conventional cost function estimations. This aggregate R&D measure appeared to reduce costs. The impact was highest in the food, chemicals, machinery, and electrical equipment industries.

27. See Harhoff (1993) for Germany; Klette (1991) for Norway.

industries and firms performing the R&D.[28] The sole exception is Hall and Mairesse, which concluded that in France the returns to R&D were 50 percent higher for those firms for which the government funded more than 20 percent of their R&D.[29] The result may reflect a combination of the relatively high prices at which the output of these industries is sold (at least to some of their customers) and the lack of good price deflators that would correct for the first problem.

How should the finding of zero returns to federal R&D at the firm or industry level be interpreted? Industries in which federal R&D is a major share of R&D spending are the following, in order of importance: guided missiles and spacecraft (376), ordnance and accessories (348), aircraft and parts (372), fabricated metal (34), transportation equipment excluding motor vehicles and aircraft (373–75, 379), communication equipment and electronic components (365–67), and electric transmission and distribution equipment (361). In some of these industries, such as ordnance and guided missiles, the government is the major purchaser, and both prices and output deflators for these industries can be expected to convey little information about true productivity. But in some of the other industries, customers for the products embodying the R&D should include other firms, and this would tend to move measured prices closer to true quality-adjusted prices.[30] A more likely explanation is simply that the R&D is not subject to a market test, and so it should not be expected to yield returns that are localized to the firms and industries that perform it or use its output.

Another channel exists through which the government funding of R&D in private firms may act as a subsidy to innovative activity: It may raise the productivity of privately funded R&D and thus cause the firms to increase their own spending. From the perspective of private or social returns to R&D, this complementarity effect ought to be included. It implies that government-funded R&D raises the returns to private R&D, which in effect lowers the cost of R&D to the firm. Several studies have tried to measure the response of private R&D to government-funded

---

28. Lichtenberg (1992).

29. Hall and Mairesse (1995). Approximately forty or fifty such firms are in the sample, and they are primarily in the machinery, electrical machinery, electronics, and aircraft industries.

30. But see the discussion in the previous section. If the price deflators make no attempts at quality adjustment, the revenue elasticity, and not the output elasticity, is being measured. Thus zero private returns to federal R&D may be found, but this may have no implication for productivity effects.

Table 6-3. *Federally Funded R&D in Industry*
Millions of 1982 dollars

Industry (2-digit SIC)	1980	1989
Chemicals (28, extracting 283)	401	67
Drugs and medicines (283)	31	3
Petroleum refining and extracting (29)	177	D[a]
Stone, clay, and glass (32)	51	D[a]
Primary metals (33)	156	27
Fabricated metals (34)	58	107
Machinery (35, extracting 357)	429	
Computing equipment (357)	326	D[a]
Communications equipment and electronics (366, 367)	4,367	4,107
Other electronics equipment (36 ex 366, 367)	D[a]	27
Motor vehicles and other (37 ex 372, 376)	661	1,694
Aircraft (372, 376)	7,732	15,544
Professional and scientific instruments (38)	669	99
Other manufacturing (20–27, 30, 31, 39)	401	924
Nonmanufacturing	907	2,150
Total	16,366	24,833

Source: National Science Foundation (1987, tables 6-5, 6-6, and 6-7); and National Science Foundation (1991, table 4-8).
a. D means the National Science Foundation omitted the number to avoid disclosing the operations of individual companies.

R&D within the firm, and all have concluded that there is a small complementarity effect, on the order of 7 percent (every dollar of federally funded R&D raises the firm's private R&D spending by seven cents).[31] Using survey data and focusing only on energy R&D, Edwin Mansfield and his colleagues found essentially the same number for a sample of forty large eastern U.S. manufacturing firms.[32] A number of this magnitude, although interesting because it is not negative, will have a relatively small effect on the measured private or social returns to this government R&D.

Table 6-3 provides a breakdown of the industries that receive federal R&D funds, drawn from the publications of the National Science Foundation. The table is incomplete owing to the spottiness with which the data have been reported in recent years, but it indicates that only two industries receive more than 75 percent of this funding: aircraft and parts, and communication equipment and electronic components. This reinforces the view that cross-industry regressions may not be the way to

31. Link (1981); Levy and Terleckyj (1982); Levin and Reiss (1984); Scott (1984).
32. Mansfield and Switzer (1984).

look for returns to this kind of R&D. Fortunately, in the case of aircraft and civilian space technology, two excellent case studies are available: Mowery on aircraft and Henry R. Hertzfeld on civilian space technology.[33] The technologies examined by these two studies are different—the first is close to commercial application in general, and the second is more like the supercollider.

Mowery's paper presents several estimates of the rates of return to the government's investment in technology applicable to commercial aircraft. It concludes that these returns depend heavily on how much of the military portion of the aircraft research budget is allocated to civilian aircraft and on whether the budget of the Civil Aeronautics Board (CAB), which had an impact on the diffusion of the technology through its regulatory role, is included. The paper's important finding is that for this technology, the successful outcome of federal R&D efforts depended on the fact that the funding extended far beyond simple basic research into diffusion and utilization. Mowery emphasizes the role of backward linkages in the technological change process in this industry.

The rate of return estimates for civilian R&D investment computed by Mowery illustrate the problems of computing returns to specific R&D investments by the government in environments where there are multiple spillovers, possibly long lags before the embodiment of the output of the investment into products, and institutional structures that impose costs on the economy. He obtains a range of estimates of the internal rate of return to investment in civilian aircraft technology all the way from − 4.3 percent to 60 percent. This range does not reflect the true range possible, because he did not compute the return under the most extreme combination of assumptions.[34] The size of the returns hinges on what one includes as the R&D cost in this industry: industry-financed R&D (always), government-financed R&D on civilian aircraft (most of the time), government-financed R&D on military aircraft (up to 25 percent), and the welfare costs of CAB regulation (from zero to 100 percent). The estimates are also affected by the choice of the lag between R&D spending and its embodiment in aircraft; a range of zero to seven years is used. Were one forced to choose a number from this menu, an estimate with a seven-year lag that accounted for the returns to all research on civilian aircraft would seem the most appropriate; this estimate was 24 percent. However, the important message in the paper is that this estimate should

33. Mowery (1985); Hertzfeld (1985).
34. These estimates are reported in table 5 of Mowery (1985).

be qualified by the regulatory environment in which the industry was operating and that potential spillovers from R&D on military aircraft have been neither measured nor accounted for.

In his review of the measurement of the economic impact of federal R&D in civilian space technologies, Hertzfeld is more cautious than Mowery, refusing even to report measures of the rate of return to such activities. This is probably appropriate when dealing with technologies with such a long and uncertain payoff, and whose goal may not be the achievement of a particular target social rate of return. Hertzfeld surveys the range of evaluation types, from large macroeconomic studies to the computation of patent statistics. The macroeconomic studies seem problematical for the obvious reason: Sorting out the benefits from a single federal R&D program from those of a hundred other programs is next to impossible using aggregate economy time-series data. The other studies fall into two basic categories: The first traces the commercial application of particular technologies and attempts to construct cost-benefit ratios. The second relies on patent statistics and measures of the licensing success of NASA. Although the second type clearly indicates a certain amount of commercial activity resulting from research done for NASA, the studies are essentially uninformative as the actual economic contribution of the patented innovations. As Hertzfeld indicates, none of these measures captures the importance of the technological advances arising from the space endeavor—satellite communications, weather and remote sensing satellites, private space launch vehicles, and new materials such as carbon and graphite composites.

Science and Industrial Innovation

Large chunks of nondefense federal R&D spending goes to health and the space program, the rest to energy (including large-scale basic physics research) and other areas, mostly basic scientific research (see table 6-2). Tracing the downstream effects of spending on basic scientific research in any area is a daunting task, and few researchers have attempted to quantify the returns to this activity. As Nathan Rosenberg has pointed out, often technological innovations are based not on new scientific discoveries, but on old science, that is, on scientific principles that have been known for decades or more.[35] Only one serious attempt has been made

35. Rosenberg (1994, pp. 142–43).

at quantifying the contribution of basic science research to productivity growth at a fairly aggregate level, that by James D. Adams and by Adams and Leo Sveikauskas.[36]

Adams describes the construction of a series of industry- and field-specific stocks of scientific knowledge based on the counts of articles published in a large number of scientific journals, weighted by the number of scientists in that field employed in the industry, either currently or with a lag of less than ten years.[37] While such stocks do not summarize investment in basic science in any dollar-financial sense, they allow certain hypotheses to be tested about the contribution of science to industry and possibly reflect on the aggregate effects of science on productivity growth in ways that are not available using the much more refined and targeted, but potentially biased, case study approach. In use, these publication counts are both field-specific and can be weighted by the number of scientists within a particular field working within an industry, so that one can characterize not only the stock of knowledge available, but also the potential to make use of it.

The drawbacks of Adams's approach were the variation in the relative importance of a scientific paper across disciplines, the imprecision with which lags are measured (citation data might help here), the arbitrariness of the obsolescence assumption used (about thirteen years), the unidimensional direction of information flow, and so forth.[38] Nevertheless it is a worthwhile addition to the arsenal for the measurement of the sources of technical change. Adams uses these stocks to help explain total factor productivity growth in a set of nineteen manufacturing industries from

36. Adams (1990, 1993); Adams and Sveikauskas (1993).

37. Adams (1990). The scientific fields considered are agriculture, biology, chemistry, computer science, engineering (combined), geology, mathematics and statistics, medicine, and physics. Typically, these articles are counted with a lag in constructing the stock of knowledge—about twenty to thirty years for most fields and ten years for engineering and computer science. The length of the lag is chosen by some preliminary data exploration for the best predictors of productivity growth. As in the case of R&D spending, estimating the lag length precisely is difficult.

38. Adams and Sveikauskas (1993) present some evidence that science precedes industrial R&D, which then leads to effects on output. But the evidence is not completely compelling, as this fact seems to have been built into their model. Although the nature of technological change may have evolved over the past century toward change that is more purposefully knowledge-based, Rosenberg's (1982) critique of the linear model for scientific research and innovation presumably still has some validity. It would be interesting in this connection to conduct model-free causality tests on the processes of scientific output, patenting, R&D investment, and the numbers of scientists and engineers in a field, in the spirit of Pakes (1985).

1953 to 1980 and finds that they enter significantly, with a longer lag of twenty years preferred for their own industry stock of knowledge and thirty years for the knowledge that spills over from other industries.[39] This work is suggestive, although fraught with interpretive difficulties because of its time-series nature. The topic could be pursued further with more detailed data in the future.

Wesley Cohen, Richard Florida, and W. Richard Goe; Adam B. Jaffe; and Edwin Mansfield all document the flow of university-based scientific research toward industry.[40] Cohen, Florida, and Goe report on the results of a major survey of the activities of University-Industry Research Centers (UIRCs) in the United States. Among other findings, they report that UIRCs have become the principal vehicle for direct industry support of academic science and engineering R&D, although government provides half of their total funding. Patent production was comparable to university patent production as a whole, indicating that, in many ways, these centers are simply an extension of the normal university-based research system. The most interesting finding was that the closer integration of industry and university research reflected in UIRC formation "appears to pose a trade-off for society" in that industrial participation promotes technical advance while it restricts communication flow and information sharing and causes publication delays.[41] The centers were successful in achieving the goal of bringing technical advances in the university lab to industry and the commercialization stage sooner.

Jaffe is closest in spirit to Adams, although he goes only as far as the technological output of university-based R&D and does not continue all the way to economic outcomes.[42] He uses variation across states in corporate and university-based R&D to tease out the contribution of university R&D to corporate patenting activity at the state level in five broad technological areas (drugs and medicine, chemicals, electronics and electrical, mechanical arts, and other). With various corrections for the simultaneity between corporate R&D and university R&D at the state level, he finds that the overall elasticity of patenting with respect to university R&D is about 0.1 and that for corporate R&D is 0.9. Because corporate R&D is six times the level of university R&D, this result implies that the marginal product in terms of patents is nearly equal. He

---

39. Adams (1990).
40. Cohen, Florida, and Goe (1994); Jaffe (1989); Mansfield (1991).
41. Cohen, Florida, and Goe (1994).
42. Jaffe (1989).

also finds that the effects are even bigger in the medical, chemical, and electrical technology areas (0.13 to 0.19). Although measuring the output of university-based R&D in this way is subject to all the usual problems associated with the use of patents as an indicator of technological output, patents arguably are as well suited for this particular exercise as they might be anywhere, because propensities to patent across states, even if they exist, are unlikely to be correlated with anything in particular (once one controls for the industrial mix).[43]

Finally, Mansfield takes more of a case study approach. He begins with commercial innovations in seventy-six firms and traces these back to their academic source (if there is one).[44] Using estimates obtained from the firms concerning the importance of recent academic research for their innovations and the time lags between that research and the first commercial introduction of the relevant product, as well as the total sales from such products, he is able to produce a rough estimate of the returns to the firms from the academic research. This is probably a lower bound on the social return, as it does not include any benefits flowing to consumers from the new products that are above the prices they have paid.

Mansfield reports that approximately 10 percent of new products and processes of these firms could not have been developed in the absence of recent academic research.[45] Using a series of heroic, but plausible, assumptions, he is able to compute a rough lower bound to the social rate of return to this academic research and reports estimates in the 20-30 percent range. As he is careful to point out, his estimates ignore the social benefits from other innovations based on the same academic research, those stemming from sales beyond the first four years of commercialization, those accruing outside the United States, and those accruing to consumers and other firms in and outside of the industry in question.

In his 1995 study, Mansfield goes back to the academic research itself, finding that about two-thirds of the funding of the researchers and projects that ultimately generated these innovations came from the federal government. This is to be compared with the overall 60 percent share of federal government funding in the research performed at universities and colleges, implying that government-funded R&D is slightly more oriented toward science that is eventually useful for industry than university R&D

---

43. See Griliches, Pakes, and Hall (1987) and Griliches (1990) for further discussion of the use of patent statistics as economic indicators.

44. Mansfield (1991, 1992, 1995).

45. Mansfield (1991).

as a whole. The conclusion from this series of studies is that academic research, much of which is funded by the federal government, is likely to generate extremely high social rates of return in spite of the difficulty of measurement of these returns.

Conclusion

In the past twenty-five or so years, both an increased understanding of the difficulties of measuring returns to R&D precisely and (in spite of this difficulty) some substantive results have been reached.

First, when the primary output of R&D is new and improved products, the allocation of the measured returns to R&D between private returns (the returns to the individual or organization undertaking the R&D) and excess social returns (the returns to society at large net of the private returns) depends crucially on the price indices that are used.

Second, given the intertemporal nature of most R&D investment projects, establishing the lag structure of the contribution of R&D to productivity and measuring the depreciation rate of R&D capital are difficult tasks. Most measures of the total returns to a particular R&D dollar will be imprecise, because these estimates depend crucially on the time pattern of the returns.

Third, the private returns to R&D in U.S. manufacturing have declined between the 1960s and 1980s, approaching something like a normal rate of return, typical of that obtained by the firms from their other activities.

Fourth, the excess private returns to federal R&D performed by individual firms is measured to be zero overall in the United States and in several other countries. Given the goals of such R&D (primarily defense, space exploration, and health in the United States, other national goals such as technological catch-up in other countries), this result is perhaps not surprising.

Fifth, case study evidence of individual research areas (such as satellites and civilian aircraft) supports the view that the social return to such R&D can be substantial, although extremely difficult to trace and measure. Little precise measurement has been made of the returns to federally funded basic science, except in a fairly aggregative manner. But, again, case study and the history of individual technologies suggest that these returns are positive and could be substantial.

What are the areas in which further research would be helpful in answering the kinds of questions asked by government policymakers? To answer this question, a brief review is necessary of the issues that policymakers might like addressed. Given the overwhelming evidence that some positive externalities exist for some types of R&D, the questions being asked typically fall into two categories: How much government subsidy should exist for R&D investment, and on what types of investment should it be spent? The set of policy instruments typically under consideration includes lowering the private cost of R&D (tax credits), direct government subsidy to private firms (for example, defense and energy) or universities (basic science), and performance by the government itself (for example, space and health).

I have written elsewhere about the issues surrounding the use of tax policy to lower the private cost of R&D, thereby increasing the level of privately supported R&D toward that called for by the social returns to such R&D.[46] In evaluating the use of tax policy to achieve this goal, the important issue is the dominant role played by the choice of price deflator in the allocation of the returns to private R&D between a firm and its customers.

The computer industry provides a clear example that the measured gap between private and social returns to R&D, which might be used to guide tax policy toward private R&D, depends crucially on whether and how the output a particular industry is adjusted for the quality change induced by the R&D. Using a conventional price index to measure output, the conclusion could be reached that the industry has achieved a social rate of return to its R&D investment that is similar to the return on ordinary capital. Using a hedonic price deflator that adjusts for the rapid quality change in this industry, the conclusion could be reached that the social return to R&D in the industry has been substantial. Simply put, the elasticity of industry sales with respect to R&D is the sum of two elasticities: the elasticity of price with respect to R&D and the elasticity of output with respect to R&D. If input costs are controlled, firms are concerned with the sales elasticity, consumers benefit from the price elasticity, and society as a whole cares about the output elasticity. Given this, how the sales growth in an industry is decomposed into price and output growth is important for the measurement of returns.

The importance of deflators in computing the returns to R&D calls for further research into the quality adjustment of output deflators at a fairly detailed industry level. A first step has been made by the Bureau of

46. Hall (1993c, 1995).

Economic Analysis in the computer industry, and some efforts also exist in autos and pharmaceuticals, but more is needed. The results should be incorporated into the National Income Accounts, as has been done with computers. In passing, the importance of extending some of this effort to outputs in the service sector should be mentioned, because an increasing share of the nation's industrial R&D is going toward this sector.

In many ways, the question about which the least is known is: What types of R&D investment should be subsidized and can the subsidy be targeted without inducing enormous rent-seeking activity? This question is important for a couple of reasons. First, whether the tax subsidy to R&D induces R&D spending that is as socially useful as direct subsidy is uncertain. Many (including myself) would argue that maximizing the diversity of ideas through decentralized choice of projects is important and would therefore favor some kind of tax subsidy approach on a priori grounds. It would be interesting, however, to attempt to track the progress of the various recent project-oriented funding programs undertaken by the Clinton administration to try to gain some insight into this question.

A related issue concerns the behavior of industry itself in choosing R&D projects. Mansfield presents the finding that two-thirds of the R&D conducted by private firms in the United States is product-oriented instead of process-oriented (as opposed to one-third in Japan).[47] Is this good or bad? Some product-oriented R&D benefits the consumer, as indicated in the computer industry example, but some may be an attempt to increase the market power of one firm at the expense of others by means of product differentiation, with no real welfare gain for consumers. Careful empirical research on this topic in several industries would be helpful in understanding the ultimate effectiveness of the R&D tax credit as innovation policy.

Second, considerable debate has ensued in the United States and other countries over the benefits of targeting government-funded research more closely to specific goals. To help evaluate the trade-off, the line of research that focuses on the downstream benefits of government investment in basic science and technology should be pursued, using the various approaches of Adams (stocks of scientific papers); Jaffe, Manuel Trajtenberg, and Rebecca Henderson (citation-weighted patents); and Lynne Zucker, Michael Darby, and Jeffrey Armstrong (biotechnology case

47. Mansfield (1988).

study, using statistical methods), among others.[48] Jaffe, Trajtenberg, and I have begun a project using patent citations that should produce the information that would enable one to link government activity in a technology to the activity of private firms. The benefit of this type of project is that, in the absence of R&D spending by field (or even in its presence), the use of bibliometric measures is both more output-oriented and more informative as to specific technological field. Research into the benefits of basic science is important because they are likely to be large and they are difficult to measure. Case study evidence, although positive, occasionally suffers from its focus on winners, and it would be helpful to try to fill it out with the computation of returns as in Mansfield's 1991 study.

Comment by Van Doorn Ooms

Edwin Mansfield summarizes primarily the findings of his own research on the contributions of new technology to the economy. For many, this would be a brief undertaking and perhaps not worth commenting upon. For Mansfield, it is the opposite. This industry, which is no longer a cottage industry, over the past nearly a quarter of a century has belonged in large part to him.

Bronwyn H. Hall entered the industry later and, as befits a later entrant, has raised some important challenges. Here she focuses primarily on the impact of government and university-based research on individual firms' contribution to overall productivity growth. In particular, she raises questions about what is known about private returns to research and development (R&D) in light of the conceptual and methodological problems involved in the research.

The two papers reach different conclusions about progress in the field. Mansfield is heartened that a great many gaps in understanding have been filled, at least partially. Hall is discouraged to discover how few of these topics have been completely explored or understood since 1972, even though a great deal of effort has been expended on research.

I am not an expert in the field, so I am not competent to say whether this research glass is half full or half empty. However, this tension is hardly unique to these particular questions. It is a characteristic of eco-

48. Adams (1990); Jaffe, Trajtenberg, and Henderson (1993); Zucker, Darby, and Armstrong (1994).

nomic research broadly and of scientific research paradigms more generally. Nevertheless, Hall is correct to call attention to some of the peculiar problems in this field. As she notes, the problems relate not only to inadequate or unsuitable data, which is characteristic of virtually all fields of economic research, but also to an inability to describe the kind of data desired in some instances—and this is not characteristic of all fields.

M. Abramovitz many years ago warned that the residual measures of technical change were measures of ignorance. The objects of inquiry are still being specified. And while some progress has been made, the difficulties remain formidable.

Mansfield provides a well-organized summary of the major issues and one that is accessible to noneconomists involved in the policy process. A few require underlining.

First, investments in new technology do produce large returns that are diffused among immediate consumers and innovators and the larger society, with a persistent gap between private and social returns. That gap provides a strong case, at least in principle, for public intervention. This is hardly news, but it raises important issues today in a political environment in which it is assumed by many, and certainly by many in the media and the public at large, that the government can do nothing right. Many large public interventions, as Linda Cohen and Roger Noll have shown, unfortunately lend support to that belief.

Second, legal and institutional protections to the innovation process, such as patents, have limited value. Here, time—and lead time, in particular—is money. Policymakers, and especially regulators, should take note.

Another point is that Mansfield's research shows that resources do matter, and in expected and systematic ways with respect to a number of issues in this area. It is not an accident or some deep mystery of eastern culture that Japanese firms successfully appropriated returns from innovations in the United States and in Europe.

Finally, academic research and the federal support underlying it are critical to innovation and economic growth. The economic theory of public goods is strongly supported by testimony on the ground from the consumers.

I do have a couple of relatively minor questions or qualifications. One is that Mansfield finds no reason that university research should not move toward closer alliances with industry. However, he also notes that the function of universities is to provide highly trained individuals, managers, scientists, and citizens. Resources are not unlimited in universities either. They have to be allocated, and the tensions between these uses are often great. Although no necessary conflict exists here, many universities are finding that the teaching function is becoming increasingly problematic with the pressures on teachers and researchers to do other things.

Second, Mansfield notes that the economic significance of new technology depends not only on the research, per se, but also on its commercialization. Not discussed is that spillovers are increasingly likely to spill across national boundaries. Firms now are increasingly global rather than national; the spillovers are likely to be appropriated hither and yon. The question arises for national policymakers: "Who is us?" U.S. consumers, for instance, benefited greatly from Japanese process innovations, even though many of the returns did not accrue to U.S. firms.

Hall's paper reiterates how difficult it is to define, let alone to accurately measure, the variables required to estimate the impact of R&D expenditures on real value added and thereby productivity. As she says, the problem is not that computers are an anomaly. Computers are anomalous because they represent the only serious attempt to develop price indices that would allow economists to try to correctly measure quantities rather than to fall back in confusion on revenues. For that reason, I would strongly subscribe to the conclusions that she and David Mowery came to on the statistical research agenda.

Despite many of the qualifications that have been noted here today, by Hall, Michael J. Boskin, and others, I think most economists believe that spillovers are large, the social returns to R&D being an order of magnitude of perhaps twice that of risky physical capital. In principle, then, a strong case can be made for public intervention to support research. Nevertheless, the political process is likely to become even less supportive than it has been. The characteristics of research are not likely to be especially attractive to elected policymakers. Research expenditures are characterized by long lags before the innovation takes place, very high uncertainty, and beneficiaries that are not easily identifiable. All of those are anathema to politicians who would prefer speed, certainty, and being able to target expenditures on powerful beneficiaries.

This is not a new problem, but a long-standing one in terms of the federal budget. Over the past thirty years, what the Office of Management and Budget has characterized as short-term benefits, principally transfer programs, have risen from about 6 percent of potential gross domestic product (GDP) to approximately 12 percent. Expenditures generally characterized as long-term investments, including R&D and physical capital, have remained at about 2 percent of GDP, or almost exactly where they were in 1960. At the margin, during the cold war, some additional resources were mobilized for long-term investments, and especially for R&D, in the name of national security. But research has now lost its security blanket, and competitiveness (if this term is viewed as a synonym for productivity) has not yet become a satisfactory replacement. It is perhaps less likely to become one while economists wage semantic war about whether the term can be used.

However, and paradoxically, as the case in principle for public support has become stronger, the difficulties in practice have increased. Congress has become an institution of independent contractors, and the discipline that was previously imposed by a stronger committee system and more careful oversight has eroded. Therefore, a coherent policy involving hard choices between expenditures, or between expenditure cuts, has become more and more difficult to implement, even if the administration proposes one. The progression is moving, with the help of armies of well-paid lobbyists, toward government by earmark, and this has been exacerbated by fiscal austerity.

As Boskin said, in principle, austerity should impose discipline and help to make hard choices. But often, at least at this stage in the development of federal budgeting, austerity largely produces what is known on Capitol Hill as fairness, which is a synonym for across-the-board cuts.

A coherent politics is badly needed—public support for long-term investments of all kinds, not only R&D, focused even more on quality and incentives than on the total amounts of resources involved; getting one is a long way off.

Comment by David C. Mowery

The Edwin Mansfield and Bronwyn H. Hall papers take different but complementary approaches, both of which should be preserved. The top-down, aggregated approach of Hall and many others who draw on public or quasi-public data and the bottom-up approach that relies on data from individual firms or individual cases, which is illustrated by the extensive labors of Mansfield, have made important contributions. However, the conclusions of these studies are limited, for a number of reasons. As retrospective studies, they tell much about how the past has operated but provide little insight into what the future will look like.

In most cases, these studies, particularly the work in the top-down tradition, do not allow one to disaggregate among industries, among types of research and development (R&D) investments, or between public and private R&D investments. The conclusions do not provide a sense of where the returns have been highest among different programmatic structures or among different areas of investment. Moreover, because one cannot determine where these returns have been highest in the past, knowing where they are likely to be highest in the future is difficult. So aggregation adds difficulties to those resulting from the retrospective focus.

The retrospective focus is problematic because of the abundant evidence that the U.S. R&D system is undergoing significant structural change. The R&D system within which many of these studies have been conducted may differ substantially from the system that will emerge over the next decade.

It is not only the U.S. R&D system that is experiencing structural change. Some of Mansfield's conclusions about U.S. versus Japanese R&D management also are likely to be qualified or slightly revised in future studies. How might these structural changes affect some of the results of Hall's and Mansfield's work? Hall suggested that some evidence is available that the expansion of R&D investment within the U.S. economy has reduced the private returns to this investment. More study of the causes and consequences of this is important for understanding what the future evolution of the system is likely to resemble.

In Mansfield's work, the role of intellectual property and the degree of formal protection afforded to industrial intellectual property by patents has substantially increased in the past fifteen years. Certainly, a number of statutory changes have been made to try and increase it; arguably, those have had some effect. Other influences are operating, but this is an important issue and may affect the speed with which new industrial technologies leak from one firm to another. Moreover, the entire approach of U.S. firms to the management of their industrial research has changed in ways that may move some practices of industrial R&D management in U.S. firms closer to what Mansfield described in Japan as an attempt to look outside the firm more systematically, to develop links to an array of external institutions, not just universities— national labs in some cases; other firms; various consortia, publicly or jointly publicly and privately financed.

So both of these analyses are snapshots of a series of rapidly evolving or changing targets and, arguably, the pace of this structural change has increased since the mid-1980s.

Two other areas may be important in Mansfield's work on the role of universities. First is the implication of greater reliance by academics on industrial funding. What does that imply about the findings of future work on the returns to academic R&D? Second, with respect to the role of public universities, how are state-level R&D or technology development and regional development strategies influencing the attractiveness of local universities for firms that are pursuing research links? A number of these programs have expanded in the past ten to fifteen years, and they are probably playing a role in some of Mansfield's findings. What that role might be is uncertain.

Other aspects of the Mansfield and Hall analyses raise broader issues. Both Mansfield and Hall emphasize the importance of adoption of the results of R&D for realization of social returns of these investments. This has some complicated implications.

As Hall and Charles L. Schultze suggested, policies favoring the adoption environment for technologies can have a substantial impact. Adoption policy that complements policy influencing investment in and the creation of much of this technology may significantly raise the social returns to these investments. As Schultze suggested, public investments in human capital may complement the innovations resulting from the R&D capital and raise the social returns to R&D investment.

Examining the effects of public R&D investment, as Hall suggested, the interaction between technology creation and technology adoption gets more complicated. Much of the postwar federal R&D spending in industry has been closely linked to federal procurement, especially in defense. The economic effects of federal R&D spending in defense-related technologies are tightly linked to the effects of federal government procurement of goods embodying many of the technologies created from the federal R&D investment.

For example, the realization of effects on industrial structure, certainly much of the so-called military-civilian spillovers, are the result of the joint influence of R&D and procurement investment. This observation has some complicated implications for the estimation of the returns to this federal R&D investment, as Frank R. Lichtenberg has explored. But federal policy on the demand side has influenced the returns to federal R&D procurement in other sectors.

In the U.S. commercial aircraft industry, for example, federal regulation of interstate air travel accelerated technology adoption. In the biomedical industry, federal policies since the mid-1960s have created a buoyant environment for the adoption of medical device technology and pharmaceutical technology by replacing the price sensitivity of users. The influence of federal policies on the adoption of innovations needs to be incorporated more directly into analysis of the effects of federal R&D programs on industrial innovation and national economic performance.

Several important areas need additional work in the aggregate approach to studying R&D. The data on R&D investment and like activities in nonmanufacturing industries must be more systematically counted, to link those data to productivity growth and to other economic quantities that matter.

But if the problems of both R&D data and deflation are severe in manufacturing, as Hall suggested, they are far more severe in the nonmanufacturing sector. Nevertheless, this is something that cannot continue to be ignored as systematically as at present. The public statistical agencies desperately need to improve data collection, analysis, and updating.

Other important research extensions are in the bottom-up analytic approach and complement the aggregate analysis of other scholars. Although individual cases can be misleading, there is a strong argument for further disaggregation and systematic differentiation among industries, among technologies, and among types of R&D investment. Reaching any conclusions about program design or evaluation without this more disaggregated research is difficult. Aggregate analysis cannot be relied upon exclusively.

Another important area for further work is a systematic attempt to update this work, much of which was written by Mansfield, to take account of some of the structural change in this R&D system, both within the United States and elsewhere in the industrial world.

Why does this matter? Enormous political pressure, which is likely to increase, is being felt for better metrics and approaches to evaluating current and prospective federal R&D programs. Current methods for assessing the economic effects of federal R&D programs are weak. As a result, economists are not well positioned to respond to the political demand for evaluation, nor are they well positioned to assess the consequences of shifts in program design or reallocation of R&D investment funds among different areas of research. One real danger is that near-term results will overshadow long-term results in such evaluation, influencing the design of programs in the area.

My final remarks deal with the role of universities in the U.S. innovation system. First is the survey, widely cited, by Wesley Cohen and Richard Florida on the role of industry R&D funding within U.S. university research. The Cohen and Florida survey revealed a surprisingly high financial contribution by industry to academic research. I do not have a good explanation for this unrepeated finding. One reason may be that the sample of universities included in the survey is a much broader one than is typically analyzed in much of the work on university-industry research relationships. The typical sample tends to be confined to the research universities. That the broader sample yields evidence of higher financial contributions from industry suggests that in overlooking this lower tier of universities, those focused on by Mansfield, a more sustained interaction may have been missed between industry and universities that goes back much earlier into the postwar period than previously thought.

Another interesting implication from Mansfield's work on the role of the second-tier universities deals with peer review. One of the subversive implications of some of Mansfield's work is that the kinds of academic excellence that are rewarded by the peer review system that has driven so much of federally funded academic research is of limited relevance to industry-funded research, at least some parts of it. What do these conclusions imply about the political robustness of peer review in a future academic research enterprise in which industry funding is likely to play a much more prominent role? The growing role of industry-funded academic research may reduce the importance of peer review in the creation or development of academic research centers.

References

Adams, James D. 1990. "Fundamental Stocks of Knowledge and Productivity Growth." *Journal of Political Economy* 98: 673–702.

———. 1993. "Science, R&D, and Invention Potential Recharge: U.S. Evidence." CES Discussion Paper 93–2. Washington: Census Bureau (January).

Adams, James D., and Leo Sveikauskas. 1993. "Academic Science, Industrial R&D, and the Growth of Inputs." CES Discussion Paper 93–1. Washington: Census Bureau (January).

Arrow, Kenneth. 1962. "Economic Welfare and the Allocation of Resources for

Invention." In *The Rate and Direction of Inventive Activity*, edited by Richard R. Nelson, 609–25. Princeton University Press.

Bartelsman, Eric J. 1990a. "Federally Sponsored R&D and Productivity Growth." FEDS 121. Washington: Federal Reserve Board of Governors.

———. 1990b. "R&D Spending and Manufacturing Productivity: An Empirical Analysis." FEDS 122. Washington: Federal Reserve Board of Governors (April).

Cohen, Wesley, Richard Florida, and W. Richard Goe. 1994. "University Industry Research Centers in the United States." Carnegie-Mellon University.

Cohen, Wesley M., and Richard C. Levin. 1989. "Empirical Studies of Innovation and Market Structure." In *The Handbook of Industrial Organization*, edited by Richard Schmalensee and Robert D. Willig, 1059–1107. Vol. II. Amsterdam: North-Holland.

David, Paul A., David Mowery, and W. Edward Steinmueller. 1988. "The Economic Analysis of Payoffs from Basic Research: An Examination of the Case of Particle Physics Research." CEPR Publication 122. Stanford University (November).

———. 1992. "Analyzing the Economic Payoffs from Basic Research." *Economics of Innovation and New Technology* 2: 73–89.

Griliches, Zvi. 1979. "Issues in Assessing the Contribution of R&D to Productivity Growth." *Bell Journal of Economics* 10: 92–116.

———. 1980. "Returns to Research and Development Expenditures in the Private Sector." In *New Developments in Productivity Measurement and Analysis*, edited by John W. Kendrick and Beatrice N. Vaccara, 419–62. University of Chicago Press.

———. 1990. "Patent Statistics as Economic Indicators: A Survey." *Journal of Economic Literature* 27: 1661–1707.

———. 1992. "The Search for R&D Spillovers." *Scandinavian Journal of Economics* 94 (3, Supplement): 529–47.

———. 1993. "Productivity and the Data Constraint." *American Economic Review* 83: 1–43.

Griliches, Zvi, and Frank Lichtenberg. 1984. "R&D and Productivity Growth at the Industry Level: Is There Still a Relationship?" In *R&D, Patents, and Productivity*, edited by Zvi Griliches, 465–501. Chicago University Press.

Griliches, Zvi, and Jacques Mairesse. 1984. "Productivity and R&D at the Firm Level." In *R&D, Patents, and Productivity*, edited by Zvi Griliches, 339–74. University of Chicago Press.

Griliches, Zvi, Ariel Pakes, and Bronwyn H. Hall. 1987. "The Value of Patents as Economic Indicators." In *Economic Policy and Technological Performance*, edited by Partha Dasgupta and Paul Stoneman. Cambridge University Press.

Hall, Bronwyn H. 1993a. "The Stock Market Valuation of R&D Investment in the 1980s." *American Economic Review* 83: 259–64.

————. 1993b. "Industrial Research during the 1980s: Did the Rate of Return Fall?" *Brookings Papers on Economic Activity: Microeconomics* (2): 289–344.

————. 1993c. "R&D Tax Policy during the Eighties: Success or Failure?" *Tax Policy and the Economy* 7: 1–36.

————. 1995. "Fiscal Policy toward R&D in the United States: Recent Experience." Paper prepared for the Organization for Economic Cooperation and Development meeting on Fiscal Policy and Innovation, Paris, France, January 19.

Hall, Bronwyn H., and Jacques Mairesse. 1995. "Exploring the Relationship between R&D and Productivity in French Manufacturing Firms." *Journal of Econometrics* 65 (1): 263–93.

Hall, Bronwyn H., Zvi Griliches, and Jerry A. Hausman. 1984. "Patents and R&D: Is There a Lag?" Working Paper 1454. Cambridge, Mass.: National Bureau of Economic Research.

Harhoff, Dietmar. 1993. *R&D and Productivity in German Manufacturing Firms.* Universitaet Mannheim Zentrum fuer Europaeische Wirtschaftsforschung.

Hertzfeld, Henry R. 1985. "Measuring the Economic Impact of Federal Research and Development Investments in Civilian Space Activities." Paper prepared for the Workshop on the Federal Role in Research and Development, National Academies of Science and Engineering, November 21–22.

Jaffe, Adam B. 1989. "Real Effects of Academic Research." *American Economic Review* 79: 957–71.

Jaffe, Adam B., Manuel Trajtenberg, and Rebecca Henderson. 1993. "Geographic Localization of Knowledge Spillovers as Evidenced by Patent Citations." *Quarterly Journal of Economics* 108: 577–98.

Klette, Tor Jacob. 1991. "On the Importance of R&D and Ownership for Productivity Growth: Evidence from Norwegian Micro-Data 1976–85." Discussion Paper 60. Oslo: Norwegian Central Bureau of Statistics (February).

Lach, Saul, and Schanderman. 1988. "Dynamics of R&D and Investment in the Scientific Sector." Working Paper. Mount Scopus, Jerusalem: Hebrew University of Jerusalem, Department of Economics.

Levin, Richard C., and Peter Reiss. 1984. "Tests of a Schumpeterian Model of R&D and Market Structure." In *R&D, Patents, and Productivity,* edited by Zvi Griliches. University of Chicago Press.

Levy, David, and Nestor Terleckyj. 1982. "Effects of Government R&D on Private R&D Investment and Productivity: Macroeconomic Evidence." Paper prepared for the Southern Economic Association.

Lichtenberg, Frank R. 1985. "Assessing the Impact of Federal Industrial R&D Expenditure on Private R&D Activity." Paper prepared for the Workshop on the Federal Role in Research and Development, National Academies of Science and Engineering, November 21–22.

————. 1992. "R&D Investment and International Productivity Differences." Working Paper 4161. Cambridge, Mass.: National Bureau of Economic Research (September).

Lichtenberg, Frank R., and Donald Siegel. 1991. "The Impact of R&D Investment on Productivity—New Evidence Using Linked R&D-LRD Data." *Economic Inquiry* 29: 203–28.

Link, Albert N. 1981. "Allocating R&D Resources: A Study of the Determinants of R&D by Character of Use." Auburn University.

Mairesse, Jacques, and Bronwyn H. Hall. 1994. "Estimating the Productivity of R&D: An Exploration of GMM Methods Using Data on French and United States Manufacturing Firms." In *International Productivity Comparisons*, edited by Karin Wagner.

Mairesse, Jacques, and Pierre Mohnen. 1995. "Research and Development and Productivity." INSEE, Paris and Universite du Quebec a Montreal (January).

Mairesse, Jacques, and Mohamed Sassenou. 1991. "R&D and Productivity: A Survey of Econometric Studies at the Firm Level." *OECD Science-Technology Review* 8: 9–44.

Mamuneas, Theofanis P., and M. Ishaq Nadiri. 1993. *Public R&D Policies and Cost Behavior of the U.S. Manufacturing Industries*. Report 93–44. New York University, C. V. Starr Center for Applied Economics, Research (November).

Mansfield, Edwin. 1984. "R&D and Innovation: Some Empirical Findings." In *R&D, Patents, and Productivity*, edited by Zvi Griliches, 127–54. University of Chicago Press.

———. 1988. "Industrial R&D in Japan and the United States: A Comparative Study." *American Economic Review* 78: 223–28.

———. 1991. "Academic Research and Industrial Innovation." *Research Policy* 20: 1–12.

———. 1992. "Academic Research and Industrial Innovation: A Further Note." *Research Policy* 21: 295–96.

———. 1995. "Academic Research Underlying Industrial Innovations: Sources, Characteristics, and Financing." *Review of Economics and Statistics* (February): 55–65.

Mansfield, Edwin, and Lome Switzer. 1984. "Effects of Federal Support on Company-Financed R&D: The Case of Energy." *Management Science* 30: 562–71.

Mansfield, Edwin, and others. 1977. *The Production and Application of New Technology*. W. W. Norton.

Mohnen, Pierre. 1992. "International R&D Spillovers in Selected OECD Countries." Montreal, Quebec: Cahiers de recherche du departement des sciences economiques de l'UQAM, no 9208 (August).

———. 1994. "The Econometric Approach to Externalities." Montreal, Quebec: Cahiers de recherche du departement des sciences economiques de l'UQAM, no 9408 (November).

Mowery, David C. 1985. "Federal Funding of R&D in Transportation: The Case of Aviation." Paper prepared for the Workshop on the Federal Role in Research and Development, National Academies of Science and Engineering, November 21–22.

National Science Foundation. 1987. *Science and Engineering Indicators 1987.* Government Printing Office.

———. 1991. *Science and Engineering Indicators 1991.* Government Printing Office.

Nelson, Richard R. 1959. "The Simple Economics of Basic Scientific Research." *Journal of Political Economy* 67: 297–306.

Pakes, Ariel. 1985. "On Patents, R&D, and the Stock Market Rate of Return." *Journal of Political Economy* 93: 390–409.

Rosenberg, Nathan. 1982. "How Exogenous Is Science?" In *Inside the Black Box,* edited by Nathan Rosenberg, 141–59. Cambridge University Press.

———. 1994. "Critical Issues in Science Policy Research." In *Exploring the Black Box,* edited by Nathan Rosenberg, 139–58. Cambridge University Press.

Scott, John T. 1984. "Firm versus Industry Variability in R&D Intensity." In *R&D, Patents, and Productivity,* edited by Zvi Griliches, 233–48. University of Chicago Press.

Terleckyj, Nestor. 1985. "Measuring Economic Effects of Federal R&D Expenditures: Recent History with Special Emphasis on Federal R&D Performed in Industry." Paper prepared for the Workshop on the Federal Role in Research and Development, National Academies of Science and Engineering. November 21–22.

Zucker, Lynne G., Michael R. Darby, and Jeff Armstrong. 1994. "Intellectual Capital and the Firm: The Technology of Geographically Localized Spillovers." Working Paper 4946. Cambridge, Mass.: National Bureau of Economic Research (December).

APPENDIX
BIBLIOGRAPHIC SUMMARY OF RECENT
RESEARCH ON TECHNOLOGICAL CHANGE

In his article at the 1971 National Science Foundation conference, Professor Edwin Mansfield enumerated questions toward which future research should be directed. This bibliography is an idiosyncratic collection of papers written since then that bear on the topics he suggested.

In some respects, it is discouraging to read over the list of topics on which Mansfield recommended further research and discover how few of them have been completely explored or understood since 1972, despite the great deal of effort that has been expended. To a great extent, this reflects the extreme complexity of the relationship between technological advance and economic welfare, as well as the resistance of some of the most important concepts to quantification. Among other problems, research in this area suffers both from inadequate or unsuitable data and the inability to describe the kind of data desired. In spite of these reservations, a considerable amount has been learned through the efforts of many researchers (to say nothing of the financial support of several branches of the National Science Foundation as well as other agencies, both government and private).

Research and Development

Improvements to the Data

R&D in Various Industries

Lichtenberg, Frank R. 1990. "Issues in Measuring Industrial R&D." *Research Policy* 19: 157–66.

Mansfield, Edwin. 1981. "Composition of R&D Expenditures: Relationship to Size of Firm, Concentration, and Innovative Output." *Review of Economics and Statistics* (November): 610–15.

Better Price Indices

Jankowski, John E., Jr. 1990. "Construction of a Price Index for Industrial R&D Output." Washington: National Science Foundation.

Mansfield, Edwin. 1987. "Price Indices for R&D Inputs: 1969–83." *Management Science* 33: 124–29.

Mansfield, Edwin, A. Romeo, and Lome Switzer. 1983. "R&D Price Indices and Real R&D Expenditures in the United States." *Research Policy*: 105–12.

Disaggregate R&D in Models Explaining Productivity Growth

Cohen, Wesley M., and David C. Mowery. 1984. *The Internal Characteristics of the Firm and the Level and Composition of Research Spending.* Interim Report to the National Science Foundation. Grant PRA 83–10664. Carnegie-Mellon University.

Basic versus Applied and Development

Griliches, Zvi. 1986. "Productivity, R&D, and Basic Research at the Firm Level in the 1970s." *American Economic Review* 76: 141–54.

Hall, Bronwyn H., and Jacques Mairesse. 1995. "Exploring the Relationship between R&D and Productivity in French Manufacturing Firms." *Journal of Econometrics* 65: 263–93.

Lichtenberg, Frank R., and Donald Siegel. 1991. "The Impact of R&D Investment on Productivity—New Evidence Using Linked R&D-LRD Data." *Economic Inquiry* 29: 203–28.

Link, Albert N. 1982. "An Analysis of the Composition of R&D Spending." *Southern Economic Journal* 49: 342–49.

Mansfield, Edwin. 1980. "Basic Research and Productivity Increase in Manufacturing." *American Economic Review* 70: 863–73.

Private versus Publicly Funded

Bartelsman, Eric J. 1990. "R&D Spending and Manufacturing Productivity: An Empirical Analysis." FEDS 122. Washington: Federal Reserve Board of Governors (April).

Griliches, Zvi. 1986. "Productivity, R&D, and Basic Research at the Firm Level in the 1970s." *American Economic Review* 76: 141–54.

Klette, Tor Jacob. 1991. "On the Importance of R&D and Ownership for Productivity Growth: Evidence from Norwegian Micro-Data 1976–85." Discussion Paper 60. Oslo: Norwegian Central Bureau of Statistics.

Lichtenberg, Frank R. 1992. "R&D Investment and International Productivity Differences." Working Paper 4161. Cambridge, Mass.: National Bureau of Economic Research (September).

Lichtenberg, Frank R., and Donald Siegel. 1991. "The Impact of R&D Investment on Productivity—New Evidence Using Linked R&D-LRD Data." *Economic Inquiry* 29: 203–28.

Profitability and Risk, Decisionmaking at Firm Level, Strategy Literature

Hall, Bronwyn H. 1993. "Industrial Research during the 1980s: Did the Rate of Return Fall?" *Brookings Papers on Economic Activity: Microeconomics* (2): 289–344.

———. 1993. "The Stock Market's Valuation of R&D Investment during the 1980s." *American Economic Review* 83: 259–64.

Economies of Scale and Scope in R&D

Scale

Cohen, Wesley M., and Richard C. Levin. 1989. "Empirical Studies of Innovation and Market Structure." In *The Handbook of Industrial Organization,* edited by Richard Schmalensee and Robert D. Willig, 1067–72. Vol. 11. Amsterdam: North-Holland.

Scope

Helfat, Constance E. 1992. "Know-How Complementarities and Knowledge Transfer within Firms: The Case of R&D." University of Pennsylvania, Wharton School.

Henderson, Rebecca, and Iain Cockburn. 1994. "Scale, Scope, and Spillovers: The Determinants of Research Productivity in the Pharmaceutical Industry." Working Paper 4466. Cambridge, Mass.: National Bureau of Economic Research (September 1993).

Translation of Basic Science into New Products and Processes

Adams, James D. 1990. "Efficient Funding of Scientific Research: An Experiment in Applied Welfare Economics." University of Florida.

———. 1993. "Science, R&D, and Invention Potential Recharge: U.S. Evidence." CES Discussion Paper 93–2. Washington: Census Bureau.

Adams, James D., and Leo Sveikauskas. 1993. "Academic Science, Industrial R&D, and the Growth of Inputs." CES Discussion Paper 93–1. Washington: Census Bureau.

Cohen, Wesley, Richard Florida, and W. Richard Goe. 1994. "University-Industry Research Centers in the United States." Carnegie-Mellon University.

Gibbons, Michael, and Ron Johnston. 1974. "The Roles of Science in Technological Innovation." *Research Policy* 3: 220–442.

Link, Albert N., and John Rees. 1990. "Firm Size, University-Based Research, and the Returns to R&D." *Small Business Economics* 2: 25–31.

Mansfield, Edwin. 1995. "Academic Research Underlying Industrial Innovations: Sources, Characteristics, and Financing." *Review of Economics and Statistics* (February): 55–65.

Mowery, David C., and Nathan Rosenberg. 1989. "The Growing Role of Science in the Innovation Process." In *Technology and the Pursuit of Economic Growth,* edited by David C. Mowery and Nathan Rosenberg, 21–34. Cambridge University Press.

Trajtenberg, Manuel. 1990. *Economic Analysis of Product Innovation: The Case of CT Scanners.* Harvard University Press.

Trajtenberg, Manuel, Rebecca Henderson, and Adam Jaffe. 1992. "Quantifying Basicness and Appropriability of Innovations with the Aid of Patent Data: A Comparison of University and Corporate Research." Paper prepared for the International Seminar on Technological Appropriation, June.

Coupling of Industrial R&D with Marketing and Production

Mansfield, Edwin. 1981. "How Economists See R&D." *Harvard Business Review* (November): 98–106.

Appropriability

Cockburn, Iain, and Zvi Griliches. 1987. "Industry Effects and Appropriability Measures in the Stock Market's Valuation of R&D and Patents." NBER Working Paper 2465.

Cohen, Wesley, and Daniel Levinthal. 1986. "The Endogeneity of Appropriability and R&D Investment." Carnegie-Mellon University.

Hanel, Petr, and Krstian Palda. 1992. "Appropriability and Public Support of R&D in Canada." Quebec: Universite de Sherbrooke.

Levin, Richard C. 1988. "Appropriability, R&D Spending, and Technological Performance." *American Economic Review* 78: 424–28.

Levin, Richard C., and others. 1987. "Appropriating the Returns from Industrial Research and Development." *Brookings Papers on Economic Activity* 3: *Microeconomics*, 783–832.

Levin, Richard C., Wesley M. Cohen, and David C. Mowery. 1985. "R&D Appropriability, Opportunity, and Market Structure: New Evidence on Some Schumpeterian Hypotheses." Carnegie-Mellon University and Yale University.

Mansfield, Edwin. 1985. "How Rapidly Does New Industrial Technology Leak Out?" *Journal of Industrial Economics* 34: 217–23.

Nelson, Richard R., and Edward N. Wolff. 1992. "Factors behind Cross-Industry Differences in Technical Progress." Columbia University.

Pakes, Ariel. 1985. "On Patents, R&D, and the Stock Market Rate of Return." *Journal of Political Economy* 93: 390–409.

Trajtenberg, Manuel, Rebecca Henderson, and Adam Jaffe. 1994. "University versus Corporate Patents: A Window on the Basicness of Inventions." CEPR Publication 372. Stanford University (January).

The Process of Technical Change

Role of R&D in Innovation

Acs, Zoltan J., and David B. Audretsch. 1988. "Innovation in Large and Small Firms: An Empirical Analysis." *American Economic Review* 78: 678–90.

Mowery, David C., and Nathan Rosenberg. 1989. "The Growing Role of Science in the Innovation Process." In *Technology and the Pursuit of Economic Growth*, edited by David C. Mowery and Nathan Rosenberg, 21–34. Cambridge University Press.

Rosenberg, Nathan. 1994. "Uncertainty and Technical Change." Paper prepared for the Conference on Growth and Development: The Economics of the Twenty-First Century, Stanford University, Center for Economic Policy Research, July.

Determinants of the Conversion of Invention into Innovation (Commercialization)

Geroski, Paul, John van Reenen, and Chris Walters. 1994. "The Demand and Supply of Knowledge: Innovations, Patents, and Cash Flow in a Panel of British Companies." London Business School and University College London. (December).

Mazzoleni, Roberto. 1993. "A Historical Analysis of the Evolution of Numerical Control of Machine Tools: Another Story of Technological Lead and Competitive Disadvantage." Stanford University.

Trajtenberg, Manuel. 1990. *Economic Analysis of Product Innovation: The Case of CT Scanners*. Harvard University Press.

Sources of Invention and Innovation, Importance of Organization Type

Henderson, Rebecca, and Iain Cockburn. 1994. "Scale, Scope, and Spillovers: The Determinants of Research Productivity in the Pharmaceutical Industry." Working Paper 4466. Cambridge, Mass.: National Bureau of Economic Research (September 1993).

Irwin, Douglas A., and Peter J. Klenow. 1994. "High Tech R&D Subsidies: Estimating the Effects of Sematech." Working Paper 4974. Cambridge, Mass.: National Bureau of Economic Research (December).

Malerba, Franco, and Salvatore Torrisi. 1992. "Internal Capabilities and External Networks in Innovative Activities: Evidence from the Software Industry." *Economics of Innovation and New Technology* 2: 49–72.

Nelson, Richard R., and Edward N. Wolff. 1992. "Factors behind Cross-Industry Differences in Technical Progress." Columbia University.

Pisano, Gary P. 1990. "The R&D Boundaries of the Firm: An Empirical Analysis." *Administrative Science Quarterly* 35: 153–76.

von Hippel, Eric. 1976. "The Dominant Role of Users in the Scientific Instrument Innovation Process." *Research Policy* 5: 212–39.

Walsh, Vivien. 1984. "Invention and Innovation in the Chemical Industry: Demand-Pull or Discovery-Push." *Research Policy* 13: 211–34.

Effect of Market Structure on an Industry's Rate of Technological Change

Cohen, Wesley M., and Richard C. Levin. 1989. "Empirical Studies of Innovation and Market Structure." In *The Handbook of Industrial Organization*, edited by Richard Schmalensee and Robert D. Willig, 1059–1107. Vol. II. Amsterdam: North-Holland.

Mowery, David C. 1983. "Innovation, Market Structure, and Government Policy in the American Semiconductor Industry." *Research Policy* 12: 183–97.

Factors Influencing the Diffusion of Innovations

Dunne, Timothy. 1991, "Technology Usage in U.S. Manufacturing Industries: New Evidence from the Survey of Manufacturing Technology." CES Discussion Paper 91–7. Washington: Census Bureau.

———. 1994. "Plant Age and Technology Use in U.S. Manufacturing Industries." *Rand Journal of Economics* 25: 488–99.

Ireland, N., and Paul Stoneman. 1983. "Technological Diffusion, Expectations, and Welfare." University of Warwick.

Karshenas, Massoud, and Paul L. Stoneman. 1993. "Rank, Stock, Order, and Epidemic Effects in the Diffusion of New Process Technologies: An Empirical Model." *Rand Journal of Economics* 24: 503–28.

Mansfield, Edwin. 1993. "The Diffusion of Flexible Manufacturing Systems in Japan, Europe, and the United States." *Management Science* 39.

———. 1989. "The Diffusion of Industrial Robots in Japan and in the United States." *Research Policy* 18: 183–92.

Nooteboom, Bart. 1993. "Adoption, Firm Size and Risk of Implementation." *Economics of Innovation and New Technology* 2: 203–16.

Stoneman, Paul, and Myung Joong Kwon. 1993. *Technology Adoption and Firm Performance.* Coventry, United Kingdom: Warwick Business School.

Stoneman, Paul. 1984. Untitled. University of Warwick.

Trajtenberg, Manuel. 1990. *Economic Analysis of Product Innovation: The Case of CT Scanners.* Harvard University Press.

Zettelmeyer, Florian, and Paul Stoneman. 1993. "Testing Alternative Models of New Product Diffusion." *Economics of Innovation and New Technology* 2: 283–308.

Spillovers from Military, Space, and Other Federal R&D to Civilian Technology

Bartelsman, Eric J. 1990. "Federally Sponsored R&D and Productivity Growth." FEDS 121. Washington: Federal Reserve Board of Governors.

David, Paul A., David Mowery, and W. Edward Steinmueller. 1991. "Analyzing the Economic Payoffs from Basic Research." Stanford University.

Lichtenberg, Frank. 1985. "Assessing the Impact of Federal Industrial R&D Expenditure on Private R&D Activity." Paper prepared for the Workshop on the Federal Role in Research and Development, National Academies of Science and Engineering, November 21–22.

Lichtenberg, Frank R., and Donald Siegel. 1991. "The Impact of R&D Investment on Productivity—New Evidence Using Linked R&D-LRD Data." *Economics Inquiry* 29: 203–28.

Link, Albert N. 1993. "Evaluating the Advanced Technology Program: A Preliminary Assessment of Economic Impacts." *International Journal of Technology Management* 8: 726–39.

Mamuneas, Theofanis P., and M. Ishaq Nadiri. 1993. "Public R&D Policies and Cost Behavior of the U.S. Manufacturing Industries." Report 93–44. New York University, C. V. Starr Center for Applied Economics Research (November).

Mansfield, Edwin, and Lome Switzer. 1984. "Effects of Federal Support on Company-Financed R&D: The Case of Energy." *Management Science* 30: 562–71.

Mowery, David C., and Nathan Rosenberg. 1989. "Postwar Federal Investment in Research and Development." In *Technology and the Pursuit of Economic Growth*, edited by David C. Mowery and Nathan Rosenberg, 123–68. Cambridge University Press.

Nadiri, M. Ishaq, and Theofanis P. Mamuneas. 1991. "The Effects of Public Infrastructure and R&D Capital on the Cost Structure and Performance of U.S. Manufacturing Industries." Working Paper 3887. Cambridge, Mass.: National Bureau of Economic Research (October).

Reiss, Peter C. 1985. "Economic Measures of the Returns to Federal R&D." Paper prepared for the Workshop on the Federal Role in Research and Development, National Academies of Science and Engineering, November 21–22.

Terleckyj, Nestor. 1985. "Measuring Economic Effects of Federal R&D Expenditures: Recent History with Special Emphasis on Federal R&D Performed in Industry." Paper prepared for the Workshop on the Federal Role in Research and Development, National Academies of Science and Engineering, November 21–22.

Economic Growth and Productivity Increase

Improved Measurement, Extended Periods

Better Measures of Output, New PPIs for the Computing Sector,
Pharmaceutical Sector, Social Welfare Effects

Bartelsman, Eric J. 1990. "R&D Spending and Manufacturing Productivity: An Empirical Analysis." FEDS 122. Washington: Federal Reserve Board of Governors (April).

Griliches, Zvi. 1993. "Productivity and the Data Constraint." *American Economic Review* 83: 1–43.

Mairesse, Jacques, and Bronwyn H. Hall. 1994. "Estimating the Productivity of R&D: An Exploration of GMM Methods Using Data on French and United

States Manufacturing Firms." In *International Productivity Comparisons*, edited by Karin Wagner.

Trajtenberg, Manuel. 1990. *Economic Analysis of Product Innovation: The Case of CT Scanners*. Harvard University Press.

Better Inputs to Productivity Studies, Engineering Data and Experience

Lichtenberg, Frank R., and Donald Siegel. 1989. "Using Linked Census R&D-LRD Data to Analyze the Effect of R&D Investment on Total Factor Productivity Growth." CES 89–2. Washington: Census Bureau.

Interrelationship among R&D, Education, Management, and Capital Formation in Economic Growth; Human Capital Measures

Cameron, Gavin. 1995. "Innovation and Economic Growth in U.K. Manufacturing." Oxford: Nuffield College (February).

Mankiw, Gregory, David Romer, and David Weil. 1992. "A Contribution to the Empirics of Growth." *Quarterly Journal of Economics*, 107 (May): 407–37.

Technical Change Embodied in Capital, Learning-by-Doing

Jorgenson, Dale W., and Kevin Stiroh. 1994. "Computers and Growth." Discussion Paper 1707. Harvard Institute of Economic Research (December).

Interindustry, Interfirm, International, and Geographical Spillovers

Mairesse, Jacques, and Pierre Mohnen. 1995. "Research and Development and Productivity." INSEE, Paris, and Universite du Quebec a Montreal (January).

Mohnen, Pierre. 1994. "The Econometric Approach to Externalities." Montreal: Cahiers de recherche du departement des sciences economiques de l'UQAM no 9408 (November).

Interindustry and Interfirm

Bernstein, Jeffrey I., and M. Ishaq Nadiri. 1988. "Interindustry R&D Spillovers, Rates of Return, and Production in High-Tech Industries." *American Economic Review* 78: 429–34.

Bresnahan, Timothy F. 1986. "Measuring Spillovers from Technical Advance: Mainframe Computers in Financial Services." *American Economic Review* 76: 741–55.

Griliches, Zvi. 1992. "The Search for R&D Spillovers." *Scandinavian Journal of Economics* 94 (3, supplement): 529–47.

Jaffe, Adam. 1986. "Technological Opportunity and Spillovers of R&D: Evidence from Firms' Patents, Profits, and Market Value." *American Economic Review* 76: 984–1001.

Mansfield, Edwin, Mark Schwartz, and Samuel Wagner. 1981. "Imitation Costs and Patents: An Empirical Study." *Economic Journal* (December): 907–18.

Geographical

Audretsch, David B., and Maryann P. Feldman. 1994. "R&D Spillovers and the Geography of Innovation and Production." Carnegie-Mellon University and Wissenschaftszentrum Berlin.

Jaffe, Adam B., Manuel Trajtenberg, and Rebecca Henderson. 1993. "Geographic Localization of Knowledge Spillovers as Evidenced by Patent Citations." *Quarterly Journal of Economics* 108: 577–98.

Zucker, Lynne G., Michael R. Darby, and Jeffrey Armstrong. 1994. "Intellectual Capital and the Firm: The Technology of Geographically Localized Spillovers." Working Paper 4946. Cambridge, Mass.: National Bureau of Economic Research (December).

International

Bernstein, Jeffrey I., and Pierre Mohnen. 1994. "International R&D Spillovers between U.S. and Japanese R&D Intensive Sectors." Working Paper 4682. Cambridge, Mass.: National Bureau of Economic Research (March).

Lichtenberg, Frank R. 1992. "R&D Investment and International Productivity Differences." Working Paper 4161. Cambridge, Mass.: National Bureau of Economic Research (September).

Mohnen, Pierre. 1990. "The Impact of Foreign R&D on Canadian Manufacturing Total Factor Productivity Growth." Montreal, Quebec: UQAM, CREPE, Cahier de recherche no 58 (July).

———. 1992. "International R&D Spillovers in Selected OECD Countries." Montreal, Quebec: UQAM, Cahiers de recherche du departement des sciences economique no 9208 (August).

International Comparisons

Blundell, Richard, Rachel Griffith, and John van Reenen. 1995. "Dynamic Count Data Models of Technological Innovation." *Economic Journal* 105 (March): 333–44.

Coe, D., and Elhanan Helpman. 1993. "International R&D Spillovers." Discussion Paper 840. London: Center for Economic Policy Research.

Hall, Bronwyn H., and Jacques Mairesse. Forthcoming. "Exploring the Relationship between R&D and Productivity in French Manufacturing." *Journal of Econometrics.*

Hanel, Petr. 1994. "R&D: Interindustry and International Spillovers of Technology and the Total Factor Productivity Growth of Manufacturing Industries in Canada, 1974–89." Sherbrooke, Quebec: U de Sherbrooke Departement d'Economique Cahier de recherche 94–04 (October).

Harhoff, Dietmar. 1993. "R&D and Productivity in German Manufacturing Firms." Mannheim, Germany: Zentrum für Europäische Wirtschaftsforschung.

———. 1995. "R&D, Spillovers, and Productivity in German Manufacturing Firms." Mannheim, Germany: Universitaet Mannheim and Zentrum für Europäische Wirtschaftsforschung (January).

Klette, Tor Jacob. 1992. "On the Importance of R&D and Ownership for Productivity Growth: Evidence from Norwegian Micro-Data 1976–85." Discussion Paper 60. Oslo: Norwegian Central Bureau of Statistics.

Lederman, Leonard L. 1993. "A Comparative Analysis of Civilian Technology Strategies among Some Nations: France, the Federal Republic of Germany, Japan, the United Kingdom, and the United States." *Policy Studies Journal* 22: 279–95.

Lichtenberg, Frank R. 1992. "R&D Investment and International Productivity Differences." Working Paper 4161. Cambridge, Mass.: National Bureau of Economic Research (September).

Mairesse, Jacques, and Mohamed Sassenou. 1991. "R&D and Productivity: A Survey of Econometric Studies at the Firm Level." *OECD Science-Technology Review* 8: 9–44.

Mansfield, Edwin. 1988. "The Speed and Cost of Industrial Innovation in Japan and the United States: External versus Internal Technology." *Management Science* 34: 1157–68.

———. 1988. "Industrial R&D in Japan and the United States: A Comparative Study." *American Economic Review* 78: 223–28.

CHAPTER 7

Quality of Life Returns from Basic Research

Susan E. Cozzens

QUALITY OF LIFE goals for science and technology have taken on new prominence in many industrialized nations in the post–cold war period. National security has by no means disappeared as a reason for maintaining national technical capacity, nor have the benefits of research for innovation and economic competitiveness been discounted. But healthy, educated citizens and high-quality, high-paying jobs have moved up in the list of priorities. And equally important, new ways of reaching the older goals are being articulated. The United States is not only concerned about being among the world's leaders in science and engineering, but also doing so in a way that lives up to America's ideals of equal opportunity and utilizes all of its talents. The aim is for a changing world, a more technologically intensive one, as well as a socially and environmentally sustainable one. The objective has shifted from simply developing technology to doing so responsibly, and the goal of wealth has been transformed into that of prosperity.

My questions here concern how to achieve what is wanted in these added dimensions of scientific progress and technological change. What has basic research contributed to the quality of life as defined today? What criteria should be used to judge whether current efforts are succeeding? What goals can be set and how can research be managed to achieve the returns in quality of life desired now and for future generations?

A crisp definition of quality of life goals eludes me, as it has eluded and probably will elude public discussion. All true national goals are fuzzy and permit many definitions, because they represent broad consensus instead of scholarly precision. National security and economic competitiveness are goals of this sort, and quality of life need not be any clearer. Like other national goals, quality of life goals are translated through decisionmaking processes into quality of life objectives. The objectives need to be proposed in context, debated publicly, and put into operation in relation to specific programs, not set by fiat in a paper such as this one.

In general, however, the term *quality of life* applies to:

—the way the full range of Americans live, including their work, family, home, and community lives;

—objectives with importance above and beyond their economic exchange value;

—how science is done and how technology is used, as well as the content of science or performance characteristics of technology.

In short, they are goals that allow Americans to articulate what they need from scientists and engineers to live in a world of their own choosing.

Two kinds of goals that fit these criteria have emerged in recent policy discussions. I refer to them as what-goals and how-goals. Quality of life what-goals for research include health, education, and environmental quality. Quality of life how-goals include equity, democracy, and community.[1] Current discussions on science policy give the impression that setting quality of life goals is sufficient to assure that they are achieved. Several reports from prestigious organizations have called for reorienting post–cold war science through priority and goal setting. But quality of life goals have been set in the past. And when political leaders have asked what has been achieved, they have been met with protestations that research cannot be held accountable in this way. When leaders have proposed a dialogue that clarifies what the country can expect from the basic research community with regard to national goals, they have met with cries of outrage and accusations that they are trying to murder science. One listener described these reactions as "somewhat hysterical."[2] Surely they underrate the demonstrable contributions of research to current quality of life and also fail to take seriously the necessity of achieving those contributions in a cost-effective way and the responsibility for monitoring the unplanned side-effects of research, both positive and negative. A more sober assessment is in order.

1. What-goals enter science policy discussion most often as program objectives. How-goals often enter as selection and evaluation criteria.
2. Mikulski (1994, p. 222).

The Conceptual Framework

Bringing People Back In

In policy discussions, research is often described abstractly: The federal government supports research; research produces benefits for society. This form of language follows the conventions of scientific writing, which call on authors to remove themselves from their texts. But it distorts the assessment of the connections between research and quality of life by leaving the people who do science out of view. In this particular bit of imagination stretching, I bring the people back in, for two reasons. First, when research is seen as the concrete activity of specific individuals, whose jobs involve activities other than research and who also lead lives outside science, then their institutions and their communities are brought back into view as well. This step is important in assessing the full connections of research with the quality of life of Americans.

Second, bringing the people back in will reorient thinking about the purposes of government support for research. Research here is treated as the advancement of knowledge in the sciences. The central goal of government support of fundamental research is considered not the production of knowledge per se, but the maintenance and renewal of national technical capacity in the form of researchers. This does not mean that the federal government should support researchers whether or not they are producing new knowledge. Instead, attention is focused on the consequences for society that come from having researchers around, consequences that go beyond the production of knowledge.

One of the most important consequences of bringing the people back in is that thinking in terms of targeting knowledge production can cease. Instead of trying to plan what needs to be known next to move toward a quality of life goal, efforts can be made to determine where, in society or the economy, to develop technical capacity. That is, the questions can be asked: Where should researchers be? Whom should they be talking to? Specifics could be left completely in their hands.

A Three-Dimensional Model

A simplifying scheme can trace the connections between researchers and quality of life outcomes. The results researchers produce can be mapped in three directions: knowledge, practice, and education.[3]

3. I have referred to this elsewhere as the keystone model. See proceedings volume from

—The primary job of the researcher is to produce knowledge that contributes to a research front. The researcher's contributions then become part of a larger pool of knowledge.

—The long-term benefits associated with research begin when someone draws on that pool of knowledge. That person might use the knowledge to improve practice, for example, in industry, medicine, or agriculture.

—The person might draw on knowledge in the context of teaching, to transmit the current best understanding of how the natural and technological worlds work to students and the general public.

Because the benefits of science come from the pool of knowledge, not directly from an individual knowledge contribution, seeing the connections between basic research and the quality of life is difficult when research is thought of only in terms of knowledge outputs.

When researchers themselves are focused on as the primary product of federal support programs, however, the other routes through which benefits are accomplished spring into view. For example, the researcher may use his or her expanding expertise and familiarity with the knowledge pool directly in practice, by consulting (as engineers do often, with public and private organizations), by combining practice and research in the same job description (as medical researchers who are also active clinicians do), or by serving as an adviser to a governmental or nongovernmental organization at national, state, or local levels. Some researchers, those who do graduate training, also contribute to practice by training professionals. When they leave their training in a research setting, professionals are up-to-date on the contents of the knowledge pool and equipped with the inclination and skills to dip into it later when they need it. In an analogous way, researchers on university campuses contribute to the utilization of the knowledge pool by providing up-to-date undergraduate teaching and creating a life-long love of learning among college-educated people. Among Americans, half go to college, and the other half are taught by people who went to college. The quality of undergraduate science education and the extent to which it creates a sense of competence with regard to technical matters thus has a very wide influence on quality of life for all Americans.

American Association for the Advancement of Science (AAAS) colloquium on science and technology, April 1994.

Approaches to Assessing Quality of Life Outcomes

Quality of life outcomes of basic research, like its economic outcomes, can be assessed either at aggregate or at program level. The logic of the assessment is different at these two levels, and the way the two sets of analyses interact is important in building up a knowledge base on quality of life returns.

At the aggregate level, the assessment begins by identifying indicators of the relevant what-goals—in the biomedical case, for example, indicators of health. Given that the goals themselves are broad, multiple indicators will undoubtedly be available, and experts will not all agree on any single key indicator. But broad agreement is likely, at least retrospectively, about the direction of progress implied in the indicators as a set.

The analysis then attributes portions of the change in the indicators to research, through several available types of logic. The timing of the research contributions and change in the indicators can be used, as, for example, in this analysis of the contribution of medical research to health outcomes:

> The tide of infectious and nutritional diseases was rapidly receding when the laboratory scientist moved into action at the end of the past century. . . . In reality, the monstrous specter of infection had become but an enfeebled shadow of its former self by the time serums, vaccines, and drugs became available to combat microbes.[4]

Alternatively, direct causal connections, for example, can be traced between the introduction of a vaccine and a subsequent dramatic drop in the incidence of a disease; or indirect causal connections, through the facilitation of an approach or technique or the incorporation of the knowledge into a technology.

A genre of research called retrospective studies has demonstrated the general usefulness of such explorations of the rich network of connections linking research with outcomes.[5] Most retrospective studies, however, have started with specific advances in an area of practice and traced

4. Rene Dubos, quoted in Brown (1979, p. 220).
5. The classic study for biomedicine is Comroe and Dripps (1976). A more recent example is Institute for Defense Analysis (1991).

the events that led to them. To judge quality of life outcomes, a broader set of indicators must be used, because some data may show lack of or the opposite of progress. Asking why some problems are getting worse and what research capacity is needed to reverse those trends is as important as assessing what research has contributed to past progress toward quality of life. Both how-goals and what-goals in such an analysis should be examined; that is, for example, not only whether biomedical research contributed ultimately to health, but also whether it did so in a way that contributed to or detracted from equity and contributed to or detracted from the empowerment of citizens.

At the program level, the analysis begins from the opposite end of the causal chains, in the activities of the researchers themselves. It tracks their immediate outputs—knowledge production, usually in the form of publications; direct contributions to practice and indirect ones through the training of professionals; and contributions to the educational stream through undergraduate teaching and general science activities. Using the set of institutional and social linkages that retrospective studies have identified, the analyst can then see where and how the researchers in the program are contributing, not in terms of direct production of outcomes, but by setting in motion the sorts of processes that are expected to produce outcomes.

Individual programs have their own goals, of both the what and how variety, and cannot be evaluated by whether they have moved the entire set of quality of life indicators in a particular direction. Federal-level leadership, however, can compare the total of indicators in a particular quality of life area with the total of contributions and consequences that result from federal programs and assess whether the portfolio of programs is connected richly enough to the institutions that produce what-outcomes and is operating in a manner that will produce how-outcomes. Thus aggregate retrospective studies combine with program assessments to give high-level decisionmakers the needed information base to manage for quality of life outcomes.

Research Contributions to Quality of Life

Not as much is known as could be about the connections between researchers and the changing quality of life in America, but some things are clear.

The One-Hundred-Year Perspective

RESEARCHERS. In biomedicine, the one-hundred-year viewpoint harkens back to an era of a major reform in medical education in America.[6] In the post–Civil War years, doctors were a weak professional group in American society, competing with each other for a small market and with no special claim to authority over medical care. For admission to medical school, only a high school education was required. The medical curriculum followed no special order, and the graduation requirements were minimal. In 1871 Harvard led the way to raising standards, extending the academic year from four to nine months and the full course from two to three years. Other institutions found they could not afford not to follow suit. In 1890 the new Association of American Medical Colleges, which included the most progressive one-third of medical schools, set minimum requirements of three years, six months a year, with laboratory work in histology, chemistry, and pathology.

In 1893 the Johns Hopkins University opened its medical school, embodying "a conception of medical education as a field of graduate study, rooted in basic science and hospital medicine, that was eventually to govern all institutions in the country. Scientific research and clinical instruction now moved to center stage."[7] Hopkins recruited faculty nationally instead of locally, required two years of preclinical sciences and two on the wards, and created advanced residencies in specialized fields. Hopkins invented partnerships in biomedicine research, establishing a hospital in connection with the medical school and joining science firmly to clinical hospital practice. Hopkins graduates exported this model around the country over the next decades.

Disparities grew between schools following this model and smaller, commercial, or special interest institutions and reached a watershed early in the twentieth century. In 1910 a young Hopkins-connected educator, Abraham Flexner, was hired by the American Medical Association to study the quality of education at American medical colleges. His report recommended closing most of them; he would have kept only 31 out of 131. In the wake of the report, many did not survive, and those that did raised their standards. An influx of private philanthropic support rein-

---

6. This section is drawn from Starr (1982). See also Stevens (1971); Rosenberg (1987); and Stevens (1989).

7. Starr (1982, p. 115).

forced this shift. The Rockefeller Foundation's General Education Board channeled $91 million into medical schools in the two decades following the publication of the Flexner Report, with seven institutions receiving more than two-thirds of the funds. Its staff actively encouraged medical education more closely linked to medical science than to medical practice. "These policies determined not so much which institutions would survive as which would dominate, how they would be run, and what ideals would prevail."[8]

KNOWLEDGE. Thus the seeds were sown for the post–World War II growth of federally sponsored biomedical research. Aside from the institutional base built through private philanthropy at a small group of research hospitals, only a few government-sponsored health-oriented research institutions existed before the Second World War. These included a small Laboratory of Hygiene that would become the National Institutes of Health.

Two important aspects of the pattern of biomedical research in the twentieth century appeared during this period. On the one hand, science within the medical school context was more likely to be linked to medical practice than science done outside; thus from the viewpoint of researchers themselves, the knowledge-practice link was forged in the Hopkins model. Assuming that research knowledge was to be useful to medical practice, the alternative was that researchers should be located outside medical institutions—clearly a second choice. On the other hand, under the science-based model as it evolved, the technical capacity developed through research, embodied in the researchers themselves and those they trained, was highly concentrated in a few institutions. As a result, from the viewpoint of medical practitioners, a gap opened up between the new scientific medicine and general medical practice.

PRACTICE. The gap took several forms. First, at the individual level, "academic and private physicians began to diverge and represent distinctive interests and values."[9] Thus, while medical science could respond to clinical problems as they appeared in the context of the great teaching hospital, most doctors might find the research results distant and hard to assimilate into their practice. In addition,

8. Starr (1982, p. 121).
9. Starr (1982, p. 122).

the medical profession grew more uniform in its social composition. The high costs of medical education and more stringent requirements limited the entry of students from the lower and working classes. And deliberate policies of discrimination against Jews, women, and blacks promoted still greater social homogeneity. The opening of medicine to immigrants and women, which the competitive system of medical education allowed in the 1890s, was now reversed.[10]

Second, at the community level, the move to scientific medicine created great disparities. Flexner's recommendations to reduce the number of medical schools to thirty-one would have left twenty states without any medical schools. In the end, more than seventy survived, and state legislatures stepped in to ensure at least one institution in virtually every state. Even these institutions did not produce doctors for every community, however. "Before the Flexner report, there had been seven medical schools for blacks in the United States; only Howard and Meharry survived" the 1910 watershed.[11] African Americans faced outright exclusion from internships and hospital privileges at most institutions. In 1930 only one out of every three thousand black Americans was a doctor, and in the Deep South, the ratio was one in more than fourteen thousand.[12] The comparable numbers for Asian and Native Americans can only be imagined. Doctors from communities that lost access to medical expertise in the wake of the Flexner report complained that, while their local medical training might not be the equivalent of Harvard's or the University of Pennsylvania's, it was better than having no doctors at all in poor and rural communities. "Would you say," one wrote, "that such people should be denied physicians? Can the wealthy who are in a minority say to the poor majority, you shall not have a doctor?"[13] The implicit answer was, "Yes."

EDUCATION. Another consequence of the way biomedical research was established institutionally was its isolation from the general educa-

10. Starr (1982, p. 124).
11. Starr (1982, p. 124). On the impact of the Flexner report at Howard, see also Manning (1983).
12. Starr (1982, p.124).
13. Starr (1982, p. 125). One might assume that European American doctors were providing adequate health care for African Americans as well, but segregation and racism must have intervened. See, for example, Maya Angelou's recounting of her experience with a white dentist in Arkansas; Angelou (1971).

tional stream. Biomedical researchers were medical school faculty and did not generally teach undergraduates, even at Harvard, Johns Hopkins, or any of the other universities where they were located. What biomedical researchers were learning about basic biology, then, had to be diffused through the general pool of biological knowledge before it could reach college students, and through them the school curriculum. The routes for diffusion of research knowledge to general medical practitioners were strong in comparison. Thus the institutional location of biomedical research contributed to a growing gap in expertise between medical practitioners and their patients.

OUTCOMES. From the advent of scientific medicine to the Second World War, the average health of Americans improved. But medicine itself, and scientific medicine in particular, is generally credited with only a small part of this improvement. A working group of the Carnegie Commission on Science, Technology, and Government gives considerable credit to factors other than research:

> The health of the American people, as judged by life expectancy, has been improving since the turn of the century. Initial improvements in longevity primarily reflected diminished mortality from infections and were largely attributable to improvements in sanitation and nutrition and to the development of effective vaccines. Sulfonamides, penicillin, and other antibiotics contributed to a further decrease in death rates.[14]

In short, specific preventive and therapeutic measures developed during this period probably only accelerated a decline in mortality rates that was already under way.

To focus only on such aggregate assessment of the products of biomedical science, however, misses half of the challenge of examining quality of life issues. What matters is not only what was done, but also how it was done. To complete this analysis, questions would have to be asked about the differential distribution of health outcomes among the American people. The continuing exclusion must be taken into consideration of many population groups from the medical profession generally and from biomedical research careers specifically in the first half of the twen-

14. Carnegie Commission (1992).

tieth century. And the relationship of patients and their families to doctors and the medical system generally also must be taken into account.

The Fifty-Year Perspective

KNOWLEDGE. Research within the medical context received a tremendous boost in the post–World War II period and continues to dominate the profile of federally sponsored basic research. Following stunning contributions to the war effort, researchers gained dominance in the Public Health Service in the late 1940s. Two wealthy women, Mary Lasker and Florence Mahoney, raised the public profile of biomedical research and stimulated a coalition among researchers, members of Congress, and segments of the mobilized public, which resulted in a phenomenal growth rate in funding in the 1950s. The new biomedical coalition adopted a strategy of seeking funds under the rubric of specific diseases, and the National Institutes of Health (NIH) subdivided rapidly on this basis. Researchers retained a remarkable degree of control over research, through institutional mechanisms such as peer review.[15]

In the content of knowledge itself, by far the most prominent trend has been the move toward the molecular level of analysis. Within the framework of disease-oriented institutes, biomedical researchers are more united in their exploration of the fundamental dynamics of genetic expression. Moving to this level of analysis has also reunited biomedical knowledge with the rest of biology more powerfully than before.

RESEARCHERS. The enormous quantitative increase in the amount of biomedical research activity has brought a broader, more complex institutional base of researchers. Three-quarters of NIH's extramural research is done in higher educational institutions; slightly more than half of the support goes to medical schools alone.[16] Outside the medical schools, large numbers of applications come from biology, chemistry, and biochemistry departments, but two-thirds are spread across the rest of the university. About 20 percent of extramural funding goes to nonprofit research institutes, independent hospitals, and other nonprofit institutions. Geographically, while the funds are not distributed evenly on a per capita basis, every state gets something; Idaho, ranked last, received more

15. This discussion is drawn from Starr, who uses in particular Shryock (1947); Strickland (1972); and Rettig (1977).
16. National Institutes of Health (1993).

than $1 million in NIH funds in 1992. About one in four NIH applications is from a person with a medical degree.[17] Industry has also built its technical capacity by hiring biomedical researchers, a trend that was only beginning in the 1930s and did not become a major activity until after the war. Pharmaceutical firms have significant in-house research efforts, and the medical device industry is also research-intensive, using researchers with many kinds of expertise, including biomedical.

Minority communities, however, continue to lack access to this technical capacity. Biomedical research is still largely a Caucasian domain in America, more Caucasian than medical practice itself. Recently, an advisory body of minority researchers and practitioners chose building minority biomedical technical capacity as the keystone of the Minority Health Initiative at the National Institutes of Health. A counterpart group considering women's health issues had focused on the shorter-term strategy of addressing the knowledge base with regard to specific female health concerns such as breast cancer; only in later rounds of resource allocation did the question of training programs for female researchers arise. But the minority group allocated more than half of the available resources from the beginning to training programs, thus affirming the perceived importance of the equitable distribution of technical capacity among Americans.[18]

PRACTICE. The connections between present-day biomedical research and various kinds of practice are rich. Biomedical researchers themselves consult, own biotechnology and drug firms, and are practicing physicians, among playing other roles. In medical schools, they train new generations of doctors and provide continuing education for doctors already in practice. Through graduate training for researchers who end up in industry and continuing collaborative relations with industrial researchers, they build the technical capacity of pharmaceutical and other firms, maintain personal connections between industrial and academic research, and increase the likelihood that industrial researchers will have the knowledge and inclination to draw on the pool of biomedical knowledge as they need to in their work. From the viewpoint of researchers, institutional locations create a powerful network keeping them aware of

17. All data through this section are from National Institutes of Health (1993).
18. See Minority Programs Fact-Finding Team Recommendations, presented to the NIH associate director for minority programs, February 1992; and National Institutes of Health (1992).

practical problems and providing channels for their research results to be incorporated into practice.

From the wider viewpoint of medical practice, however, the network may not appear so effective. A large proportion of physicians are learning the basic sciences at a high level in their medical curriculum and are therefore prepared as well as possible for the struggle to keep up with research results that appear after they are in practice. As the pool of biomedical knowledge spreads to become a lake and then an ocean, however, the challenge of continuing education for physicians looms large. The physician's most common information on research results may be the media and drug salesmen. As doctors move through their three decades or so of practice, they inevitably find themselves further and further from the research front, incorporating new therapeutic approaches with more difficulty and less understanding into their work.

Some of the most controversial outcomes of the growth of biomedical research are mediated through the training of medical practitioners and the symbiotic relationship between science-based medicine and medical technology. A growing body of anthropological evidence is available examining the relationship between patients and the medical system as a factor shaping a personal sense of autonomy and empowerment. In this analysis, the growing scientific knowledge of medical practitioners can increase a sense of powerlessness in patients. This sense is exacerbated by medical technology, which transports one's experience of one's own body outside the body, onto imaging screens and into numerical test results. The women's health movement has been particularly vocal in drawing attention to the disempowerment that can result from standard doctor-patient or patient-hospital interactions.[19] The point applies with even more force to poor communities, whether rural or urban. Given the number of interactions Americans have over a lifetime with medical care providers, any such sense of disempowerment can accumulate to the point of leading to a larger sense of alienation from authority. The situation is thus of no inconsiderable consequence for general levels of democratic participation.

EDUCATION. One clear benefit from the vast growth of biomedical knowledge and the push to the molecular level is the increased probability of integrating biomedical results with those of the rest of biology. Biomedical research is no longer confined to the medical schools within

19. See, for example, Fee (1983); Martin (1987); Nelkin and Tanredi (1989).

the university. Biology and chemistry teachers, who regularly teach undergraduates, are also contributing to the biomedical effort. Thus the basic results of biomedical research, not surprisingly, are making their way into college textbooks, and thus on to school-level science training. The limitations on this diffusion route come largely from the problems of the educational system itself. The gap between what biomedical researchers know about the human body and what high school students in the nation's weakest schools know is widening. Likewise, other parts of the public are differentially exposed to new concepts in medicine. Most of modern genetics, for example, has been discovered in the time since President Bill Clinton's generation took high school biology—in the mid-1960s. Public broadcasting documentaries, science sections of newspapers, and science museums combined do not reach more than a small segment of that generation with the simplest of knowledge of the major themes of contemporary biomedical research. Physicians and other health care providers can end up being the major science educators, but in relation to specific conditions and at times of stress. Thus the physician-patient relationship takes on yet another significant role in shaping the character of life for Americans.

OUTCOMES. An overall assessment of the outcomes of post–World War II biomedical research is a matter for debate and discussion. The high cost of health care is an issue: In the 1980s health care costs rose twice as fast as general inflation.[20] And whether the gains in health were related to those costs, or to the investment in biomedical research, is controversial.

In recent years, increases in life expectancy have resulted primarily from reductions in cardiovascular death rates from stroke and coronary artery disease. These improvements reflect control of hypertension, a decrease in the prevalence of smoking, decreases in the intake of fats and cholesterol, better weight control, and healthier lifestyles. There have also been substantial reductions in death rates from certain types of cancer, owing to improvements in surgery, radiation, and chemotherapy. Unfortunately, increases in lung cancer due to smoking have approximately canceled out the successes with other forms of cancer.[21]

20. U.S. Department of Health and Human Services (1993, p. 1).
21. Carnegie Commission (1992, p. 42).

The aggregate numbers again mask differences in outcomes for different groups in the population. Between 1980 and 1990, the overall life expectancy at birth for Americans increased from 73.7 to 75.4 years. But life expectancy for the Caucasian population increased by 1.7 years and in the African American population by 1.0 year, thereby widening the gap between the two. In 1990 the infant mortality rate for Americans as a group was 9.2 deaths per 1,000 live births, but it was 7.6 for Caucasians and 18 for African Americans—and again the gap had widened between 1980 and 1990.[22]

The Next Twenty Years

The issue of the relationship of biomedical research to quality of life has been raised through several strategies. By using the perspectives of one hundred years and fifty years, the issue can be examined in terms of long-term relationships built up through institutional structures and patterns of interaction shaped by education, not only in terms of research results and their incorporation into medical practice. By focusing on researchers instead of research knowledge, attention is concentrated on institutions and the distribution of technical capacity among Americans and American communities. By using health statistics, on the one hand, consideration was given to the larger set of practices through which the benefits of biomedical research must reach the public, in particular, about the distributive aspects of the medical care system. On the other hand, by talking about health, and not only health research, the specific agendas and accomplishments of particular research communities have been put into a broader perspective of life in America. The resulting discussion bears a certain resemblance to the aggregate assessment of the contribution of basic research to the economy.

As with aggregate economic assessment, however, such broad-brush assessment of quality of life issues provides little direction to researchers, research mangers, or policymakers in how to achieve quality of life goals. The conventional tools of research policy and management—goal setting and program evaluation and planning—are equal to the task of determining how to attain these goals, although they are not being used wisely. Quality of life goals put one new requirement on research policy: the inclusion of the American public—or to be more precise, American

---

22. U.S. Department of Health and Human Services (1993, p. 1).

publics—in the processes. Just as people were brought back into the concept of what federal support for research is about, people must be brought back into research management. Researchers themselves are not the experts on quality of life outcomes; they are partners in producing them. All the partners need to be involved in running the firm.

Goal Setting

In current science policy discussions, the variety and insistence of calls to set goals for research has been striking. The Carnegie Commission report stresses the need to set long-term goals and not squander technical resources on short-term objectives with no vision of the future in mind.

> If this emphasis continues, the problems we have encountered in recent years, such as erosion of the nation's industrial competitiveness and the difficulties of meeting increasingly challenging standards of environmental quality, could overwhelm promising opportunities for progress. However, we believe there is an alternative. The United States could base its S&T [science and technology] policies more firmly on long-range considerations and link these policies to societal goals through more comprehensive assessment of opportunities, costs, and benefits.[23]

The report gives a long list of examples of major societal goals to which science and technology contribute, including personal and public health and safety, creation and maintenance of civic culture, and environmental quality and protection. It proposes a National Forum on Science and Technology Goals to facilitate "a balanced and effective interaction . . . between the scientific and engineering communities and those representing a broad range of other societal interests."[24]

The call for goals has also been voiced within government. Senator Barbara Mikulski, D-Md., from her position as chair of the Senate Appropriations Committee subcommittee with authority over the National Science Foundation (NSF), in 1993 called for 60 percent of the NSF budget to be devoted to strategic research, in areas such as high performance computing, biotechnology, materials science, and manufactur-

---

23. Carnegie Commission (1992, p. 11).
24. Carnegie Commission (1992, p. 13).

ing.[25] The National Science and Technology Council (NSTC) has taken up the challenge in its recent guidance for the budget process, which calls on agencies to justify their programs with regard to goals such as a healthy, educated citizenry, job creation and economic growth, and world leadership in science, mathematics, and engineering.[26]

For biomedical research, a goals document of approximately the kind envisioned in the Carnegie Commission report exists: *Healthy People 2000: National Health Promotion and Disease Prevention Objectives.* Billed as a "statement of national opportunities," the report is the product of a national effort, involving twenty-two expert working groups, a consortium that grew to include almost three hundred national organizations and all state health departments, and the Institute of Medicine (IOM) of the National Academy of Sciences. IOM helped the Department of Health and Human Services (HHS) convene regional and national meetings and receive testimony from more than 750 individuals and organizations in producing a draft set of objectives. After extensive review and comment, involving more than ten thousand people, the objectives were revised and refined to produce the final report.[27] The process of producing the report would seem to be a model for the sort of national forum the Carnegie Commission recommends. While probably too dominated by professional groups to meet the participatory criteria of some who wish to democratize science and technology policy, it was dramatically more open and participatory than the normal program planning and budgeting process of any research institution—even NIH, where a third of each institute's governing council are nonresearchers representing a broader view of the context of the research.

Still, *Healthy People 2000* illustrates the current lack of connection between research program planning and long-term national goal setting. The Carnegie Commission notes the problem:

> There is a mismatch between the long-term societal goals necessary for our society's well-being in the 21st century and many of the present scientific goals of research. The implications for biomedical research of a new social goal of cost-effective and equitable health

25. See "Senate Turns Up the Heat on NSF," Science 261, September 17, 1993, pp. 1512–13.

26. See "Memo Backs Basic Research with Words, Nor Cash," Science 264, June 3, 1994, pp. 1395–96.

27. The preceding sentences are taken almost verbatim from the conference edition of the report, dated September 1990.

care delivery to the entire U.S. population have not yet been carefully analyzed.[28]

The last version of the NIH strategic plan, announced to the press in the summer of 1992 but never released, seems not to have taken *Healthy People* and its national goals into account. The current HHS strategic plan focuses on programs that can produce short-term outcomes and pays scant attention to research. Comparing the preventive goals of individual NIH Institutes with those articulated in *Healthy People,* some overlaps and some divergences appear. No iterative process has linked the visionary, broadly negotiated goal statement with plans for programs at NIH.

Program Planning and Evaluation

The question of how to attain the quality of life goals is answered in the process of program planning. In theory, this process translates broad federal goals and objectives into specific program activities and budget requests. In practice, this process should be engaging agency program planning much more powerfully with national goal setting, because of two recent sets of directives:

—The NSTC budget guidance, which asks for justification of programs in terms of national goals. This request should make it easier for the NSTC to see its full portfolio of programs in relation to various goals and allow choices that give the best chance of moving toward them.

—The Government Performance and Results Act of 1993 (P.L. 103–62), which requires agencies to write strategic plans, set performance goals for the programs based on those plans, and report annually on performance.

The language of these documents tends to raise fears among scientists that policymakers are asking them to plan research. These fears are not dispelled by the accompanying disclaimers under titles such as "Voyages of Discovery" and terms such as *the spirit of pure discovery.*[29] Given the goal of a research program of putting active researchers in certain kinds of contexts, however, the reason for the fears disappears. When people are brought back in, program planning for basic research programs no longer involves controlling the contents of research projects, but instead

28. Carnegie Commission (1992, p. 43).
29. Carnegie Commission (1992, p. 10); Clinton and Gore (1994, p. 5).

pays attention to where researchers are working, whom they talk to, what communities they empower, and what else they do besides research.

Exactly what these criteria entail will vary from program to program. NSF's Engineering Research Centers and Science and Technology Centers serve as an example of how this approach can be used. The centers are not told in detail what to study, but they are required to show that they are embedded in a set of partnerships that is likely to move their research results into practice, and they are required to make a commitment to undergraduate as well as graduate education. When these broader functions of research become the focus, thinking in terms of aggregates of researchers instead of individual ones is important; no individual researcher needs to perform equally on all three dimensions of evaluation, but a set of researchers should. Thus the sum of center-affiliated researchers maintain the connections in three directions: knowledge, practice, and education. Likewise, any federal research program can be seen as a sum of researchers, who together contribute to the knowledge base, maintain connections to practice, and enhance education.

The difference between a center and a program, in this comparison, is that the center is required to have a strategic planning process that directs research toward topics that the specific clients of the center find important. Researchers and clients together participate in that process. A program consisting of a portfolio of individual grants (for individuals or teams) does not need to determine centrally what topics the investigators study. But it does need processes in which researchers and their partners together discuss how the total research ongoing in the program relates to its primary contexts of use, so that researchers themselves remain aware of these contexts and can take them into consideration in their own choice of research topics. The process of program planning provides one such opportunity, but the investigators themselves are seldom involved. Program evaluation processes, however, which usually involve extensive interaction with program participants, offer a better setting for this negotiation.[30]

30. See *Evaluation of Fundamental Research Programs: A Review of the Issues*, a report on discussion in the Practitioners' Working Group, August 1994. Background paper for "Metrics of Fundamental Science," undertaken for the National Science and Technology Council, Committee on Fundamental Science, Subcommittee on Research, by the Critical Technologies Institute, RAND. The paper will be included in the Critical Technologies Institute documentation report on the project.

Program evaluation for a program aimed at technical capacity asks not "Have the researchers in this program answered a specific question?" nor "Have they achieved some particular result?" but

—Are they doing excellent research?

—What are they learning from it, and who else is learning it?

—Who draws on the pool of knowledge these researchers contribute to?

—Are these researchers talking to the people who use that pool of knowledge, to remain aware of the long-term practical problems it relates to?

—Are these researchers empowering citizens and consumers, directly or indirectly, with the knowledge they produce and how they produce it?

The information gathered in the evaluation can be used to shape the program, for example, by shedding light on the mix of resources investigators need to do excellent research and facilitate its use, or by identifying new partners or ways of interacting with partners that can increase the program's effectiveness. The evaluation process itself raises the awareness of both researchers and partners of how the program works to contribute to quality of life goals; that is, how it sets processes in motion that bring those results. In an untargeted program, the process of evaluation thus serves some of the same functions as the strategic planning process does for a centralized research unit, but in a way that leaves choice of research topics and judgment of technical excellence entirely in the hands of researchers.

To achieve quality of life goals through such programs, however, the key is who is involved in program evaluation and planning. To see the whole set of linkages through which the program is having its impact, evaluation panels need only a minority of researchers and a wide range of people with other kinds of knowledge. One important set of such people is usually called next-stage users of research. These are people with knowledge of practice applications (such as physicians, nurses, and counselors in biomedical research). The other important set of participants are end-users—those who have experienced and thoughtfully considered the ultimate results of the researchers' work. In the biomedical case, the entire American public are potentially end-users. Just as with the selection of researchers for evaluation committees, the challenge is to find citizens with broad enough perspective and relevant experience to cover the range the committee needs. Finding such people is worth the effort, however. Unless the processes of program evaluation and planning include a wide range of such partners, they will not be able to judge quality of life outcomes effectively.

Conclusion

Does the framework apply outside biomedicine? Parts of it, in particular the link between researchers and practice goals, are well developed in military research and development (R&D). Next-stage users are involved in many planning and evaluation processes for defense research and in industry-oriented programs. End-stage users are involved in a few, including some military R&D assessment processes and NIH's councils. In disciplinary science programs, a different mix of the three dimensions is appropriate, with more emphasis on educational responsibilities, more diffuse relationships to practice, and more involvement of users within the science community, drawn both from the discipline supported by the program and from other disciplines that depend on its knowledge pool. Any federal program of research can be seen profitably within the framework, however, as long as the appropriate weight is put on each of the three dimensions.

Two related bodies of knowledge and experience need to be built up extensively, to put this framework into practice effectively. First, much more must be known about the activities and institutions that link basic research to quality of life outcomes. These are identifiable in specific cases, but no general body of knowledge, equivalent to the body of knowledge about industrial innovation processes, exists to provide general concepts in this area. Without general concepts, every retrospective study and evaluation process starts de novo to identify these links, and the overall effort required is much greater. A small community of researchers devoted to understanding these links needs to be developed, through workshops and research projects, to play a role for quality of life outcomes that parallels the role of economists of R&D in relation to industrial innovation.

Second, experience must be built up by locating appropriate public members of evaluation and planning panels, and utilizing their insights effectively. The Carnegie Commission's call for including American publics in the National Forum on Science and Technology Goals has highlighted the fact that the research policy community does not know whom to recruit for this role. NIH has traditionally looked to organized patient and patient-family groups as members of its councils, as well as nonscience professionals who bring different perspectives to the process. These are positive steps and can provide the beginnings of creative thought about broader representation. Other areas of research have scarcely begun to think about the organizational equivalents for their areas of education and practice.

Once representatives of various publics are on these panels, however, chairs and executive secretaries can easily be at a loss as to how to give them an effective voice. In the absence of a set of guidelines for effective public-researcher interaction in research decisionmaking processes, such committees are open to unconscious domination by the research community, a process that defeats the purpose of public representation. If the goal of achieving quality of life returns from basic research is serious, the benefit of the expertise of public members of these groups cannot be lost. Concerted attempts should be made to collate the experience to date with mixed committees of this sort and build a base of research knowledge that can provide practical guidance to chairs and executive secretaries of future committees.

A great deal of work needs to be done to use the framework to increase the effectiveness of programs with regard to goals that will affect the quality of life. The challenges involve both ways of bringing people back in.

—Putting the researcher back at the center of research evaluation and planning concepts. The focus too long and too exclusively has been on the knowledge products of research. Thinking about researchers as multidimensional contributors will take some imagination, and then some effort in translating that understanding into appropriate program structures and resources.

—Putting the public back in evaluation and planning processes. Public involvement in research management has been seen for too long as a threat to autonomy and a form of political control. In the framework outlined, the opposite is true. Public involvement in program evaluation and planning may be the only route, under current circumstances, to leaving researchers in control of their research topics and processes. Furthermore, it is a necessary condition for using technical capacity to contribute to quality of life goals, because researchers themselves have part of the expertise, but not all the expertise, needed to identify the complex links between research and these outcomes.

In short, a quiet revolution is needed in science policy, a transformation within existing structures and processes that will maintain, use, and expand the important strengths of the current U.S. basic research system.

Comment by Shirley M. Malcom

I enjoyed Susan E. Cozzens's paper. While the issues that she raised were provocative, in a way she did not go far enough. That might surprise her because, in many audiences, she may be criticized or looked upon as going too far. The question that she raised about the social contract is a real one. The social contract contains considerable fine print; and the devil is in the details. An interesting point was the extent to which a preoccupation with either/or issues exists. For example, I came to know the story of Johns Hopkins as an institution from reading Margaret Rossiter on women scientists in America. The founders of the Johns Hopkins Medical School ran out of money before they were able to get it set up, and they turned to a woman, heiress to the B&O fortune, to help them out. She essentially said, "I will help you raise the money if the medical school will be open to women as well as to men." The founders refused her gift the first time because they did not like the strings attached. They tried to raise the money from other sources; they could not and had to come back to her and to accept the conditions that went with the money. And so Johns Hopkins Medical School was open to women, as well as to men, from the beginning because the philanthropist insisted on that requirement from the outset.

I learned about Abraham Flexner from Ken Mannings's book on E. E. Just. Manning describes the kinds of trials and tribulations that Just went through to get his own research supported and the kind of dominating presence that the whole Flexner project had on the medical enterprise and on the pattern of support for research. A trade-off seemed to exist between quality on the one hand and access on the other. Just was a top-flight researcher in a minority institution who could not get his research supported, largely because of his race.

What would the world look like if, instead of aiming for doing away with everything that is not quality or focusing only on a certain set of institutions, people had said, "Raise the quality of a number of institutions, including those that basically are concerned with providing equity and access." If research money were poured into those institutions that guarantee equity and access, what would have been the effect on the nation's research effort? Instead, the worst of two worlds resulted: No requirements were made to guarantee access on quality research institutions, and the other institutions were removed from competition for research funds. On the one hand, access was denied, and on the other, research was restricted. Efforts may only now be catching up with this kind of preoccupation with either/or ways of doing business.

Cozzens also raises the issue about how research gets to people. The answer she provides is that research gets to people through people. If that is the case, then how the flow of scientific research takes place, and the implications for how science is taught and research performed, must be of high concern. The new view argues for a better balance between the amount of education by research universities and the amount of research done by teaching institutions. This again reflects an either/or mentality. One thing will be done in one set of institutions and another primarily in another set of institutions. If the twain meets, it is purely coincidental.

Cozzens asks how much of the gap—in terms of health conditions in society—is attributable to research. I would ask how much of the gap is attributable to lack of research. The population of researchers is homogeneous and looks different from those most underserved by present patterns of health research and delivery systems.

Current discussions about education in the sciences focus not only on transmitting knowledge, but also on the need for people to understand the nature of science. The nature of science is understood by engaging in the processes by which knowledge is created. So this leads back to interest in how research is done. It also leads back to concerns about the undergraduate part of the educational system.

Other issues remain. One is the extent to which science or knowledge from science can be gotten on demand, when people need it (just-in-time science). A disconnection exists among the different pieces of the educational and research system, despite attempts through programs to remedy the situation. For example, the new National Science Foundation Career Development Program asks that people reconnect their research, teaching, and other functions within the institutions. However, the flow of knowledge is still one way—from the researcher and the scientist to the public. The assumption is not that somehow the interaction with students and the public will give something back to knowledge and research.

A strong requirement for outreach to the public does not exist right now. One is penalized if one spends too much time reaching out to the public. The claim is that knowledge flows through the system from people to people, but the approach to the activity works in the opposite direction. Creating processes is valued more than transmitting processes.

Cozzens puts forth the possible solution of having more popular participation within the whole process of evaluation. Once the specific quality issues are addressed, then choices can be made from among what is already there through a mechanism that involves popular input.

I would say that, having served on some of those kinds of committees, the people who come from the public representation side get marginalized quickly. Once everybody else has made the decisions, these public representatives are supposed to come in and bless the decisions or else to keep quiet. It is probably a much more effective mechanism if one has technical capacity within the committee that is broadly representative of the different scientific communities, and that one comes into those meeting in both a public and technical capacity. Then proposals can be challenged on the front end.

I have seen this happen, for example, in the codevelopment of assistive technology with regard to scientists and engineers working with scientists and engineers with disabilities and working with the disability community. The scientists and engineers with disabilities can play a two-way translational role and also help to keep people honest.

The system presents many challenges, which call for the kind of reinvention that Cozzens described. But more cleverness is required as well as a recognition of what has happened in the past when the public has been given a voice in technical arguments. The result is the tyranny of experts on the technical side who fail to recognize the expertise on the people's side, on the quality of life issues. Quality of life expertise deserves equal respect but often ends up unequal in those discussions.

References

Angelou, Maya. 1971. *I Know Why the Caged Bird Sings.* Bantam Books.
Brown, E. Richard. 1979. *Rockefeller Medicine Men.* University of California Press.
Carnegie Commission on Science, Technology, and Government. 1992. *Enabling the Future: Linking Science and Technology to Societal Goals.* (September).
Clinton, William J., and Albert Gore, Jr. 1994. *Science in the National Interest.* Executive Office of the President, Office of Science and Technology Policy.
Comroe, Julius H., and R. D. Dripps. 1976. "Scientific Basis for the Support of Medical Science." *Science* 192: 105–11.
Fee, Elizabeth, ed. 1983. *Women and Health: The Politics of Sex in Medicine.* Farmindale, N.Y.: Baywood Publishing Company.
Institute for Defense Analysis. 1991. "DARPA Technical Achievements." Arlington, Va. (July).

Manning, Kenneth R. 1983. *Black Apollo of Science: The Life of Ernest Everett Just.* Oxford University Press.

Martin, Emily. 1987. *The Woman in the Body.* Beacon Press.

Mikulski, Barbara. 1994. "Science in the National Interest." *Science* 264 (April 8): 221–22.

National Institutes of Health. 1992. *Report of the National Institutes of Health: Opportunities for Research on Women's Health, Summary Report.* NIH Publication 92–3457A. Office of Research on Women's Health, Office of the Director (September).

———. 1993. *NIH Extramural Trends, Fiscal Years 1983–92.* NIH Publication 93–3506 (November).

Nelkin, Dorothy, and Laurence Tancredi. 1989. *Dangerous Diagnostics: The Social Power of Biological Information.* Basic Books.

Rettig, Richard A. 1977. *Cancer Crusade: The Story of the National Cancer Act of 1971.* Princeton University Press.

Rosenberg, Charles E. 1987. *The Care of Strangers: The Rise of America's Hospital System.* Basic Books.

Shryock, Richard. 1947. *American Medical Research.* New York: Commonwealth Fund.

Starr, Paul. 1982. *The Social Transformation of American Medicine.* Basic Books.

Stevens, Rosemary. 1971. *American Medicine and the Public Interest.* Yale University Press.

———. 1989. *In Sickness and in Wealth: American Hospitals in the Twentieth Century.* Basic Books.

Strickland, Stephen. 1972. *Politics, Science, and Dread Disease: A Short History of Medical Research Policy.* Harvard University Press.

U.S. Department of Health and Human Services, National Center for Health Statistics, Centers for Disease Control and Prevention, Public Health Service. *Health United States 1992.* DHHS Publication PHS 93–1232.

Contributors

CLAUDE E. BARFIELD is a resident fellow at the American Enterprise Institute for Public Policy Research and the director of its Trade and Technology Policy Studies program. Mr. Barfield was a consultant with the Office of the U.S. Trade Representative, where he wrote the Reagan administration's "Statement of Trade Policy" in 1983. He was codirector of the staff of the President's Commission for a National Agenda for the Eighties. Mr. Barfield also served as deputy assistant secretary for research and demonstration at the Department of Housing and Urban Development. He is the author and editor of several books and articles on trade, technology, and industrial policy, including, most recently, *Capital Markets and Trade: The United States Faces a United Europe* and *Expanding U.S.-Asian Trade and Investment: New Challenges and Policy Options.*

MICHAEL J. BOSKIN is Tully M. Friedman professor of economics at Stanford Univeristy, senior fellow at the Hoover Institution, and adjunct scholar at the American Enterprise Institute for Public Policy Research. In addition, Mr. Boskin is a research associate at the National Bureau of Economic Research. From 1989 to 1993, he served as chairman of the Council of Economic Advisers (CEA), where he participated in the formulation of fiscal, trade, and regulatory policy. Before chairing the CEA, Mr. Boskin taught at Stanford University, the University of California at Berkeley, Harvard University, and Yale University. He is a frequent contributor to various magazines and newspapers as well as television.

HARVEY BROOKS is Benjamin Peirce professor of technology and public policy emeritus in the John F. Kennedy School of Government and the Gordon McKay professor of applied physics emeritus in the Division of Applied Sciences at Harvard University. Mr. Brooks first joined Harvard University in 1940 as a junior fellow and remained until 1946, when he became the associate head of the Knolls Atomic Power Laboratory for General Electric. He returned to Harvard in 1950 as Gordon McKay professor. From 1957 to 1975, Mr. Brooks served as the dean of the Division of Engineers and Applied Physics. He has been a frequent consultant to the Office of Technology Assessment of the U.S. Congress and, on science and technology policy, to the Organization for Economic Cooperation and Development and to UNESCO. His research has been in the fields of solid state physics, nuclear engineering, underwater acoustics, and, more recently, science and public policy. Mr. Brooks is the author of *The Government of Science*.

SUSAN E. COZZENS is an associate professor in the Department of Science and Technology studies at Rensselaer Polytechnic Institute and the director of graduate studies for the department. From 1981 to 1986, Ms. Cozzens was a policy analyst at the National Science Foundation (NSF), where she also served as the associate executive secretary of the Director's Advisory Committee on Merit Review. In addition, she was a consultant in the review and reorganization of the NSF's program evaluation activities. Ms. Cozzens has served on advisory committees for the American Association for the Advancement of Science, the National Academy of Sciences, and the Office of Technology Assessment of the U.S. Congress. She is the author of several books, including *Societal Control and Multiple Discovery in Science: The Opiate Receptor.*

BRONWYN H. HALL is an associate professor of economics at the University of California at Berkeley and a research associate at the National Bureau of Economic Research. Before joining the two faculties in 1987 and 1988, respectively, Ms. Hall headed a computer software firm that she continues to own. She serves on the Census Advisory Committee of the American Economic Association and is a member of the editorial board of *Economics of Innovation and New Technology*. Ms. Hall has written numerous papers, including *Industrial Research during the 1980s: Did the Rate of Return Fall?* Most recently, she was the coauthor of "Exploring the Relations between R&D and Productivity in French Manufacturing Firms," forthcoming in the *Journal of Econometrics*.

LAWRENCE J. LAU is Kwoh-Ting Li professor of economic development in the Department of Economics at Stanford University. Mr. Lau joined the Stanford faculty in 1966 as an acting assistant professor and was promoted to assistant professor, associate professor, and, in 1976, professor of economics. Mr. Lau served as the vice chairman of the Department of Economics from 1990 to 1992 and, since 1992, has been the codirector of the Asia/Pacific Research Center. Mr. Lau has been a consultant to the Department of Energy, the Department of State, the Federal Reserve Board, and the World Bank.

SHIRLEY M. MALCOM is the head of the Directorate for Education and Human Resources Programs of the American Association for the Advancement of Science. Ms. Malcom is a former high school teacher and university professor. She serves on a number of advisory committees relating to the evaluation of educational reform.

EDWIN MANSFIELD is a professor of economics and the director of the Center for Economics and Technology at the University of Pennsylvania. Before joining the University of Pennsylvania faculty, Mr. Mansfield taught at Carnegie-Mellon, the California Institute of Technology, Yale University, and Harvard University. He has been a consultant to a number of industrial firms and government agencies and has been a member of the Advisory Committee on Policy Research of the National Science Foundation, the Advisory Committee of the U.S. Bureau of the Census, and the American Association for the Advancement of Science's Committee on Science, Engineering, and Public Policy. Mr. Mansfield served as the U.S. chairman of the U.S.-U.S.S.R. Working Party on the Economics of Science and Technology and was the first U.S. economist to be invited to visit and lecture in the People's Republic of China under the 1979 Sino–American agreement.

ERNEST J. MONIZ is a professor of physics and the head of the Department of Physics at the Massachusetts Institute of Technology (MIT). From 1983 to 1991, Mr. Moniz was the director of the Bates Linear Accelerator Center at MIT. Mr. Moniz's principal area of research is in theoretical nuclear physics. He is the chairman of the Department of Energy of the National Science Foundation Nuclear Science Advisory Committee, a director on the board of American Science and Engineering, and a consultant to the Theoretical Division of the Los Alamos Scientific Laboratory.

DAVID C. MOWERY is a professor of business and public policy at the Walter A. Haas School of Business at the University of California at Berkeley and serves as the deputy director of the Consortium on Competitiveness and Cooperation, a multiuniversity research alliance focusing on research on technology management and U.S. competitiveness. Mr. Mowery taught at Carnegie-Mellon University, served as the study director for the Panel on Technology and Employment of the National Academy of Sciences, and was an international affairs fellow at the Office of the United States Trade Representative. His research concentrates on the economics of technological innovation and the effects of public policies on innovation. He has advised the Organization for Economic Cooperation and Development, various federal agencies, and industrial firms. Mr. Mowery is the author of several books, including *Science and Technology Policy in Interdependent Economies.*

VAN DOORN OOMS is the senior vice president and director of research of the Committee for Economic Development. He was executive director for policy and chief economist of the Committee on the Budget of the U.S. House of Representatives from 1989 to 1990 and chief economist on the committee staff from 1981 to 1988. Mr. Ooms was assistant director for economic policy at the Office of Management and Budget and chief economist of the Committee on the Budget of the U.S. Senate. Before coming to Washington, D.C., in 1978, Mr. Ooms taught economics at Yale University and Swarthmore College.

PAUL ROMER is a professor of economics at the University of California at Berkeley and an associate at the Canadian Institute for Advanced Research. He was a professor of economics at the University of Chicago from 1988 to 1990 and an assistant professor at the University of Rochester from 1983 to 1988. Mr. Romer was a visitor at the Center for Advanced Study in the Behavioral Sciences. His research focuses on economic growth and development.

CHARLES L. SCHULTZE is a senior fellow in the Economics Studies Program at the Brookings Institution. He served as the program's director from 1987 to 1990 and, since 1967, as senior fellow. During the Carter administration, Mr. Schultze was chairman of the Council of Economic Advisers. Before joining Brookings, Mr. Schultze was assistant director and then director of the Bureau of the Budget. Among his recent publications are *Setting Domestic Priorities: What Can Government Do?* (with

Henry Aaron; Brookings, 1992) and *Memos to the President: A Guide through Macroeconomics for the Busy Policymaker* (Brookings, 1992).

BRUCE L. R. SMITH is a senior staff member in the Center for Public Policy Education of the Brookings Institution. Before joining Brookings in 1980, Mr. Smith served as the director of the Policy Assessment Staff, Bureau of Oceans, at the Department of State. He was a professor of public law and government at Columbia University from 1966 to 1979. He served as panel coordinator for the Commission on Critical Choices for America chaired by Nelson Rockefeller. He has been a consultant to the Office of the U.S. Special Trade Representative, the Securities and Exchange Commission, and other government agencies. From 1964 to 1966, he was a senior staff member of the RAND Corporation and a lecturer at the University of California at Los Angeles. Mr. Smith's most recent books are *The Advisers: Scientists in the Policy Process* (Brookings, 1992) and *American Science Policy since World War II* (Brookings, 1990).

Index